Personnel

AWS Committee on Definitions and Symbols

G. E. Metzger, Chairman	Wright-Patterson AFB
W. L. Green, 1st Vice Chairman	Ohio State University
E. A. Harwart, 2nd Vice Chairman	Consultant
E. J. Seel, Secretary	American Welding Society
J. T. Biskup	Canadian Welding Bureau
W. F. Brown	Westinghouse Hanford
C. D. Burnham*	Consultant
R. J. Christoffel*	General Electric Company
G. B. Coates	General Electric Company
J. E. Greer	Moraine Valley Community College
M. J. Grycko	Packer Engineering Associates
A. R. Hollins, Jr.	Duke Power Company
M. J. Houle*	National Board of Boiler and Pressure Vessel Inspectors
S. R. Morse*	Deere and Company
L. C. Northard	Tennessee Valley Authority
D. H. Orts*	Armco, Incorporated
J. G. Roberts	Southern California Drafting Services
M. W. Roth**	Hobart Brothers Company
J. J. Stanczak	Steel Detailers and Designers
J. J. Vagi	Babcock and Wilcox

AWS Subcommitee on Symbols

W. L. Green, Chairman	Ohio State University
J. T. Biskup	Canadian Welding Bureau
C. D. Burnham*	Consultant
G. B. Coates	General Electric Company
M. D. Cooper***	Hobart School of Welding Technology
E. A. Harwart	Consultant
A. R. Hollins, Jr	Duke Power Company
S. R. Morse*	Deere and Company
J. G. Roberts	Southern California Drafting Services
M. W. Roth**	Hobart Brothers Company
J. J. Stanczak	Steel Detailers and Designers

*Advisory Member
**Resigned February 1985
***Appointed May 1985

Foreword

(This Foreword is not a part of A2.4-86, Standard Symbols for Welding, Brazing and Nondestructive Examination but is included for information purposes only.)

Welding cannot take its proper place as an engineering tool unless means are provided for conveying the information from the designer to the workers. Such practices as writing "to be welded throughout" or "to be completely welded" on a drawing, in effect, transfer the design responsibility from the designer to the welder, who canot be expected to know what strength is necessary. In addition to being dangerous, this practice may also be costly; certain shops, in their desire to be safe, use much more welding than is necessary.

These symbols provide the means for placing complete welding information on drawings. The system for symbolic representation of welds on engineering drawings used in this standard is consistent with the "third angle" method of projection. This is the method predominately used in the United States. In practice, many companies will need only a few of the symbols and, if they desire, can select only parts of the system that fit their needs. If this is done universally, all will be speaking the same language even though some use but a few of the symbols contained herein.

In the past, the use of the words, "far side" and "near side" in the intrepretation of welding symbols has led to confusion because when joints are shown in section, all welds are equally distant from the reader and the words "near" and "far" are meaningless. In the present system, the joint is the basis of reference. Any welded joint indicated by a symbol will always have an "arrow side" and an "other side". Accordingly, the terms *arrow side*, *other side*, and *both sides* are used herein to locate the weld with respect to the joint.

The tail of the symbol is used for designating the welding and cutting processes, as well as the welding specifications, procedures, or the supplementary information to be used in making the weld. When only the size and type of weld are specified, the information necessary for making that weld is limited. The process, identification of filler metal that is to be used, whether peening, root gouging, or other operations are required, and other pertinent data, should be known. The notation to be placed in the tail of the symbol indicating these data will usually be established by each user. If notations are not used, the tail of the symbol may be omitted.

Symbols in this publication are intended to be used to facilitate communications among designer, shop, and fabrication personnel. The usual limitations included in specifications and codes are beyond the scope of this standard.

Illustrations used with the text are intended only to show how correct applications of symbols may be used to convey welding or examination information and do not necessarily represent recommended welding or design practice.

Part B, Brazing Symbols, uses the same symbols for brazing that are used for welding.

Part C, Nondestructive Examination Symbols, establishes symbols to be used on drawings to specify nondestructive examination for determining the soundness of materials. The nondestructive examination symbols included in the standard represent nondestructive methods as discussed in the latest edition of AWS publication B1.0, *Guide for the Nondestructive Inspection of Welds*. Definitions and details for use of the various nondestructive inspection methods are found in B1.0.

A2.4 came into existence in 1976 as the result of combining and superceding two earlier documents, A2.0, *Standard Welding Symbols*, and A2.2, *Nondestructive Testing Symbols*. Both of the earlier documents had their origins in work done jointly by the American Welding Society and ASA Sectional Committee Y32. A2.0 was first published in 1947 and revised in 1958 and 1968; A2.2 first appeared in 1958 and was revised in 1969.

AWS A2.4-76, *Symbols for Welding and Nondestructive Testing*, was the first version of the combined effort prepared by the AWS Committee on Definitions and Symbols. It was revised in 1979 as A2.4-79, *Symbols for Welding and Nondestructive Testing, Including Brazing*. The present document, AWS A2.4-86, has been more accurately titled, *Standard Symbols for Welding, Brazing and Nondestructive Examination*.

The new revision provides a new symbol and rules for conveying stud welding information.

There are also many clarifications in terminology and illustrations, and an important change in organization of the material designed to facilitate use of the document.

Users of this standard are invited to suggest additional symbols or revisions for consideration by the Committee for future revision and reissue. Correspondence relating to this standard should be addressed to the Secretary, Committee on Definitions and Symbols, American Welding Society, 550 N.W. LeJeune Road, P.O. Box 351040, Miami, Florida 33135.

Key Words — Weld symbols, welding symbols, nondestructive examination symbols

**ANSI/AWS A2.4-86
An American National Standard**

Approved by
American National Standards Institute
May 20, 1986

Standard Symbols for Welding, Brazing and Nondestructive Examination

Superseding A2.4-76, A2.4-79,
A2.2-69, and A2.0-68

Prepared by
AWS Committee on Definitions and Symbols

Issued, 1986

Under the Direction of
AWS Technical Activities Committee

Approved by
AWS Board of Directors, May 20, 1986

Abstract

AWS A2.4 establishes a method of conveying instructions to the welder or brazer by means of symbols which may have several parts. Detailed instructions and examples are provided so that the welding or brazing symbol may be constructed and interpreted to cover most welded or brazed designs. Also included is a system of symbols for informing the nondestructive examination technician as to the method, frequency and extent of examination required.

AMERICAN WELDING SOCIETY
550 N.W. LeJeune Road, P.O. Box 351040, Miami, FL 33135

Policy Statement on Use of AWS Standards

All standards of the American Welding Society (codes, specifications, recommended practices, methods, etc.) are voluntary consensus standards that have been developed in accordance with the rules of the American National Standards Institute. When AWS standards are either incorporated in, or made part of, documents that are included in federal or state laws and regulations, or the regulations of other governmental bodies, their provisions carry the full legal authority of the statute. In such cases, any changes in those AWS standards must be approved by the governmental body having statutory jurisdiction before they can become a part of those laws and regulations. In all cases, these standards carry the full legal authority of the contract or other document that invokes the AWS standards. Where this contractual relationship exists, changes in or deviations from requirements of an AWS standard must be by agreement between the contracting parties.

International Standard Book Number: 0-87171-266-0

American Welding Society, 550 N.W. LeJeune Road, P. O. Box 351040, Miami, Florida 33135

© 1986 by American Welding Society. All rights reserved
Printed in the United States of America
2nd reprint, May 1990. Errata included.

Note: By publishing this standard, the American Welding Society does not insure anyone using the information it contains against any liability arising from that use. Publication of a standard by the American Welding Society does not carry with it any right to make, use, or sell any patented items. Users of the information in this standard should make an independent investigation of the validity of that information for their particular use and of the patent status of any items referred to herein.

This standard is subject to revision at any time by the Committee on Definitions and Symbols. It must be reviewed every five years and if not revised, it must be either reapproved or withdrawn. Comments (recommendations, additions, or deletions) and any pertinent data that may be of use in improving this standard are requested and should be addressed to AWS Headquarters. Such comments will receive careful consideration by the Committee on Definitions and Symbols and the author of the comments will be informed of the committee's response to the comments. Guests are invited to attend all meetings of the Committee on Definitions and Symbols to express their comments verbally. Procedures for appeal of an adverse decision concerning all such comments are provided in the Rules of Operation of the Technical Activities Committee. A copy of these Rules can be obtained from the American Welding Society, 550 N.W. LeJeune Road, P.O. Box 351040, Miami, Florida 33135.

Table of Contents

page no.

Personnel iii

Foreword iv

List of Tables viii

List of Figures ix

Part A — Welding Symbols 1

1. Basic Symbols 1
    1.1 Distinction Between Weld Symbol and Welding Symbol 1
    1.2 Illustrations 1
    1.3 Basic Weld Symbols 1
    1.4 Supplementary Symbols 1
    1.5 Standard Location of Elements of a Welding Symbol 1
    1.6 Placement of Welding Symbol 1

2. Basic Types of Joints 1

3. General Provisions 1
    3.1 Location Significance of Arrow 1
    3.2 Location of Weld With Respect to Joint 3
    3.3 Orientation of Specific Weld Symbols 6
    3.4 Break in Arrow 6
    3.5 Combined Weld Symbols 6
    3.6 Multiple Reference Lines 6
    3.7 Field Weld Symbol 9
    3.8 Extent of Welding Denoted by Symbols 9
    3.9 Weld-All-Around Symbol 14
    3.10 Tail of Welding Symbol 14
    3.11 Contours Obtained by Welding 14
    3.12 Finishing of Welds 14
    3.13 Melt-Through Symbol 15
    3.14 Melt-Through With Flange Welds 15
    3.15 Method of Drawing Symbols 15
    3.16 U.S. Customary and Metric Units 15

4. Groove Welds 15
    4.1 General 15
    4.2 Depth of Preparation and Groove Weld Size of Groove Welds 15
    4.3 Groove Dimensions 19
    4.4 Contours and Finishing of Groove Welds 24
    4.5 Back and Backing Welds 31
    4.6 Joint with Backing 31
    4.7 Joint with Spacer 34
    4.8 Consumable Inserts 34
    4.9 Groove Welds with Back Gouging 34
    4.10 Seal Welds 34
    4.11 Skewed Joints 34

5. Fillet Welds 34
    5.1 General 34
    5.2 Size of Fillet Welds 34
    5.3 Length of Fillet Welds 36
    5.4 Intermittent Fillet Welds 36
    5.5 Fillet Welds in Holes and Slots 39
    5.6 Contours and Finishing of Fillet Welds 39
    5.7 Skewed Joints 39

6. Plug Welds 39
    6.1 General 39
    6.2 Size of Plug Welds 42
    6.3 Angle of Countersink 42
    6.4 Depth of Filling 42
    6.5 Spacing of Plug Welds 42
    6.6 Contours and Finishing of Plug Welds 42
    6.7 Joints Involving Three or More Members 42

7. Slot Welds 45
    7.1 General 45
    7.2 Depth of Filling 45
    7.3 Details of Slot Welds 45
    7.4 Contours and Finishing of Slot Welds 45

8. Spot Welds 45
    8.1 General 45
    8.2 Size and Strength of Spot Welds 48
    8.3 Spacing of Spot Welds 48
    8.4 Number of Spot Welds 48
    8.5 Extent of Spot Welding 48
    8.6 Contours and Finishing of Spot Welds 50
    8.7 Multiple-Joint Spot Welds 53

9. Seam Welds 53
    9.1 General 53
    9.2 Size and Strength of Seam Welds 53
    9.3 Length of Seam Welds 53
    9.4 Intermittent Seam Welds 57
    9.5 Orientation of Seam Welding 57
    9.6 Contours and Finishing of Seam Welds 57
    9.7 Multiple-Joint Seam Welds 57

10. *Flange Welds*	57
10.1 General	57
10.2 Dimensions of Flange Welds	59
10.3 Multiple-Joint Flange Welds	62
11. *Stud Welds*	62
11.1 Side Significance	62
11.2 Stud Size	62
11.3 Spacing of Stud Welds	62
11.4 Number of Stud Welds	62
11.5 Dimension Locations	62
11.6 Location of First and Last Stud Weld	62
12. *Surfacing Welds*	62
12.1 Use of Surfacing Weld Symbol	62
12.2 Size (Thickness) of Surfacing Welds	62
12.3 Extent, Location, and Orientation of Surfacing Welds	65
12.4 Surfacing of a Weld	65
12.5 Surfacing to Adjust Dimensions	65
Part B— Brazing Symbols	65
13. *Brazed Joints*	65
Part C— Nondestructive Examination Symbols	69
14. *Elements of Nondestructive Examination Symbol*	69
14.1 Examination Method Letter Designations	69
14.2 Supplementary Symbols	69
14.3 Standard Location of Elements of Nondestructive Examination Symbol	69
15. *General Provisions*	69
15.1 Location Significance of Arrow	69
15.2 Location of Letter Designations	69
15.3 U.S. Customary and Metric Units	71
16. *Supplementary Symbols*	71
16.1 Examine-All-Around	71
16.2 Field Examinations	71
16.3 Radiation Direction	71
17. *Specifications, Codes, and References*	71
18. *Location, Orientation, and Extent of Nondestructive Examination*	71
18.1 Specifying Length of Section to be Examined	71
18.2 Number of Examinations	72
18.3 Examination of Areas	72

Appendices:

A.	Design of Standard Symbols in inches	76, 78, 80
AM.	Design of Standard Symbols in millimeters	77, 79, 81
B.	Document List	83

List of Tables

Table   **page no.**

1 — Designation of Welding and Allied Processes by Letters 73
2 — Alphabetical Cross Reference to Table 1 74
3 — Suffixes for Optional Use in Applying Welding and Allied Processes 75

List of Figures

Figure		page no.
1	Basic Weld Symbols	2
2	Supplementary Symbols	2
3	Standard Location of Elements of a Welding Symbol	3
4	Basic Joints	4
5	Application of Arrow Side and Other Side Convention	5
6	Application of Break in Arrow of Welding Symbol	7
7	Combination of Weld Symbols	7, 8
8	Designation of Location and Extent of Fillet Welds	10
9	Designation of Extent of Welding	11–13
10	Applications of Melt-Through Symbol	16
11	Designation of Groove Weld Size; Depth of Preparation Not Specified	17
12	Application of Dimensions to Groove Weld Symbols	18
13	Examples of Different Relationships Between Depth of Preparation "S" and Groove Weld Size "(E)"	20
14	Designation of Groove Weld Size With Specified Depth of Preparation	21
15	Designation of Groove Weld Size Without Specified Depth of Preparation	22
16	Combined Groove and Fillet Welds With Specified Groove Weld Size and Fillet Weld Size, and Depth of Preparation	23
17	Complete Joint Penetration Required, Joint Preparation Optional	24
18	Partial Joint Penetration Specified, Joint Preparation Optional	25
19	Application of Flare-Bevel and Flare-V-Groove Weld Symbols	26, 27
20	Designation of Root Opening of Groove Welds	28
21	Designation of Groove Angle of Groove Welds	29
22	Application of Flush and Convex Contour Symbols to Groove Weld Symbols	30
23	Application of Back or Backing Weld Symbol	32
24	Joints with Backing and Spacers	33
25	Groove Welds with Back Gouging	35, 36
26	Application of Dimensions to Fillet Weld Symbols	37
27	Application of Dimensions to Intermittent Fillet Weld Symbols	38
28	Application of Fillet Weld Symbols	40
29	Application of Plug Weld Symbol	41
30	Application of Dimensions to Plug Weld Symbols	43, 44
31	Application of Slot Weld Symbol	46
32	Application of Dimensions to Slot Weld Symbols	47
33	Application of Spot Weld Symbol	49
34	Application of Projection Weld Symbol	50
35	Application of Dimensions to Spot Weld Symbols	51, 52
36	Application of Seam Weld Symbol	54
37	Application of Dimensions to Seam Weld Symbols	55, 56
38	Application of Edge-Flange Weld Symbol	58
39	Application of Corner-Flange Weld Symbol	60
40	Application of Edge and Corner-Flange Weld Symbols	61
41	Application of Stud Weld Symbols	63
42	Application of Surfacing Weld Symbol	64
43	Application of Brazing Symbols	66–68
44	Standard Location of Elements	70

Symbols for Welding, Brazing and Nondestructive Examination

Note: This revision contains a large number of changes in content and has been very extensively reorganized to present a more logical arrangement of material. For these reasons, item number comparisons to the 1979 edition are not possible and changes have not been marked.

Part A
Welding Symbols

1. Basic Symbols

1.1 Distinction Between Weld Symbol and Welding Symbol. This standard makes a distinction between the terms *weld symbol* and *welding symbol*. The weld symbol (Figure 1) indicates the type of weld. The welding symbol is a method of representing the weld on drawings. It includes supplementary information and consists of the following eight elements. Not all elements need be used unless required for clarity.

(1) Reference line
(2) Arrow
(3) Basic weld symbols
(4) Dimensions and other data
(5) Supplementary symbols
(6) Finish symbols
(7) Tail
(8) Specification, process or other reference

1.2 Illustrations. Examples given, including dimensions, are illustrative only and are intended to facilitate communications. They are not intended to represent design practices, or to replace code or specification requirements.

1.3 Basic Weld Symbols. Basic weld symbols shall be as shown in Figure 1. The symbols shall be drawn "on" the reference line (shown dotted).

1.4 Supplementary Symbols. Supplementary symbols to be used in connection with welding symbols shall be as shown in Figure 2.

1.5 Standard Location of Elements of a Welding Symbol. The elements of a welding symbol shall have standard locations with respect to each other as shown in Figure 3. Specification and process references should be shown in the tail of the welding symbol.

1.6 Placement of Welding Symbol. The arrow of the welding symbol shall point to a line on the drawing which conclusively identifies the proposed joint. It is recommended that the arrow point to a solid line (object line, visible line); however, the arrow may point to a dashed line (invisible, hidden line).

2. Basic Types of Joints

The basic types of joints are shown in Figure 4.

3. General Provisions

3.1 Location Significance of Arrow

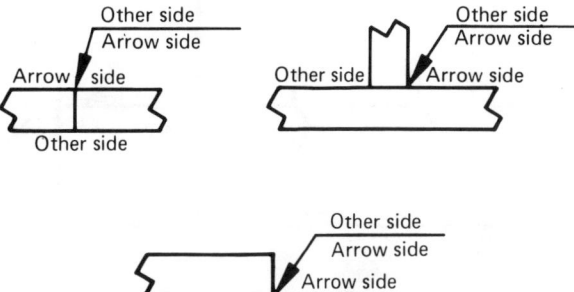

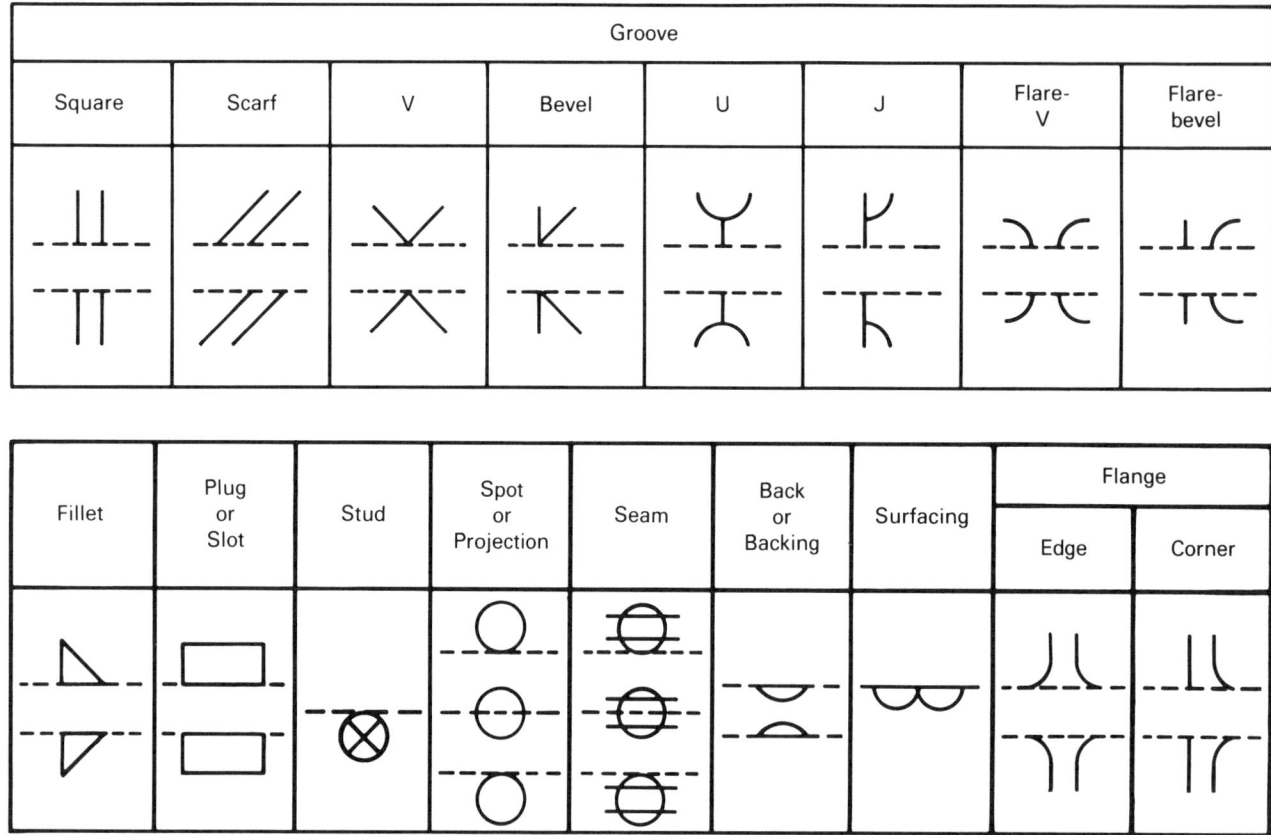

Figure 1 — Basic Weld Symbols

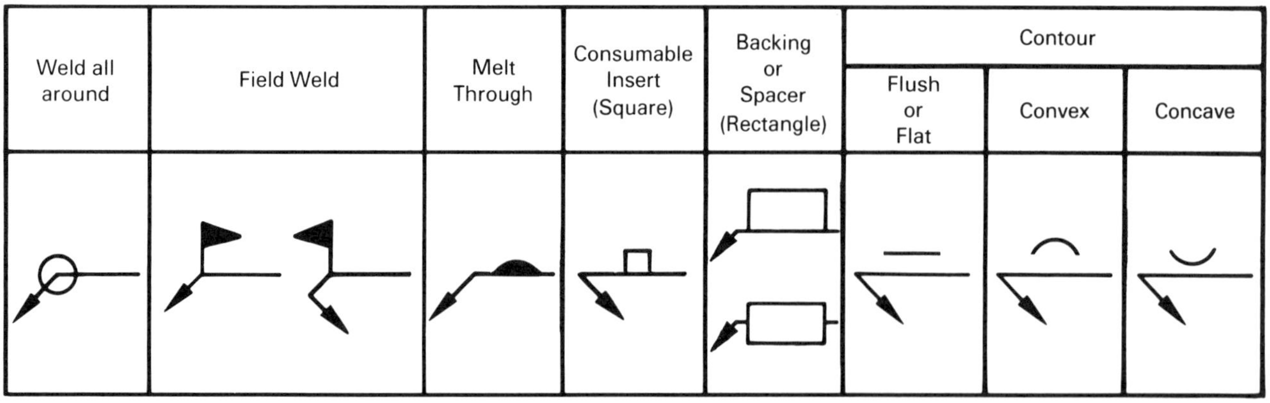

Figure 2 — Supplementary Symbols

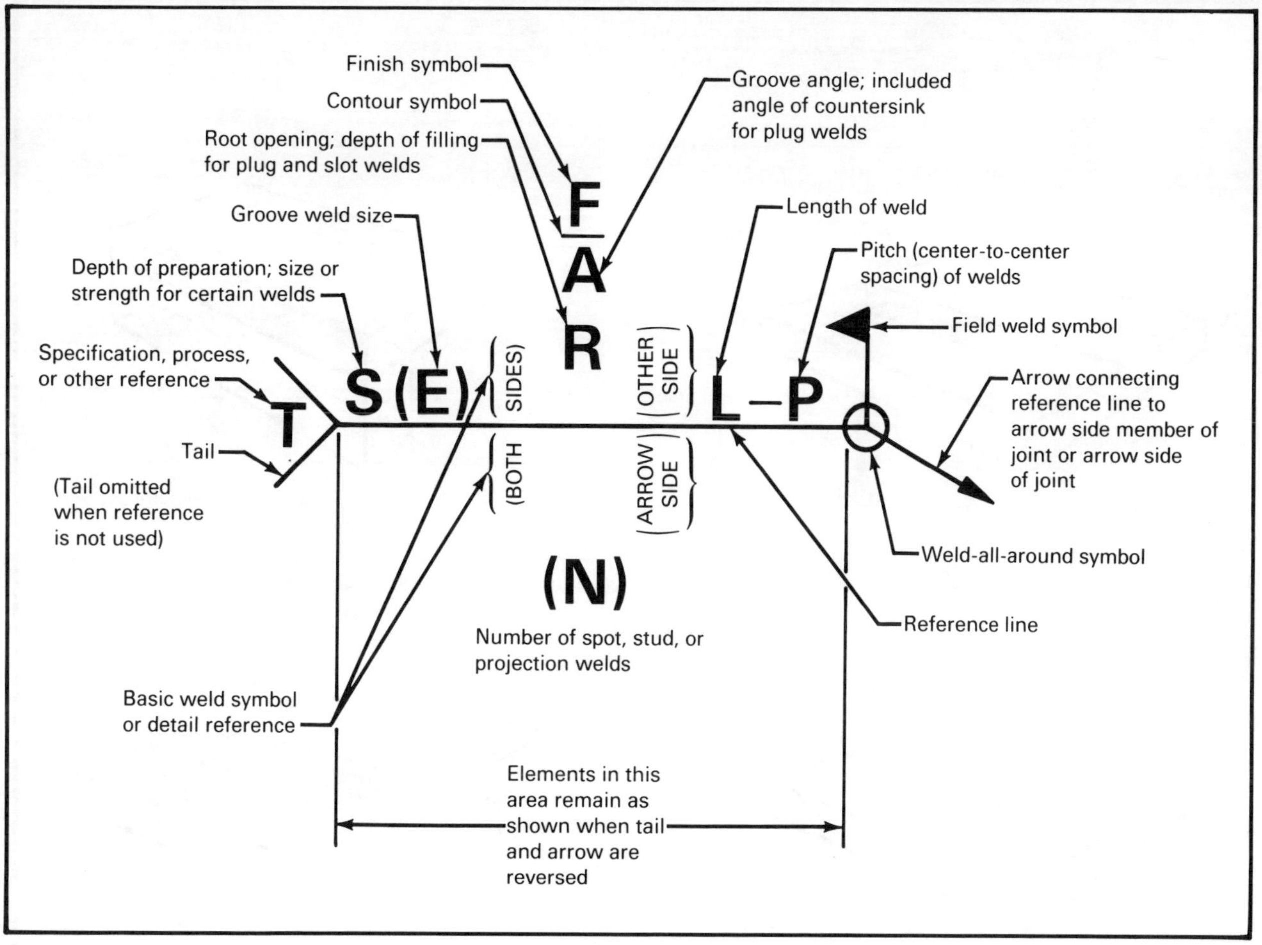

Figure 3 — Standard Location of Elements of a Welding Symbol

3.1.1 Fillet, Groove, and Flange Weld Symbols. For these symbols, the arrow shall connect the welding symbol reference line to one side of the joint, and this side shall be considered the arrow side of the joint. The side opposite the arrow side of the joint shall be considered the other side of the joint (see Figure 5).

3.1.2 Plug, Slot, Spot, Projection and Seam Weld Symbols. For these symbols, the arrow shall connect the weld symbol reference line to the outer surface of one of the joint members at the center line of the desired weld. The member toward which the arrow points shall be considered the arrow side member. The other joint member shall be considered the other side member (see Figures cited in sections 6 to 9 inclusive).

3.1.3 Symbols with No Side Significance. Some weld symbols have no arrow side or other side significance, although supplementary symbols used in conjunction with them may have such significance (see 8.1.2, 8.1.4, and Tables 1 and 2).

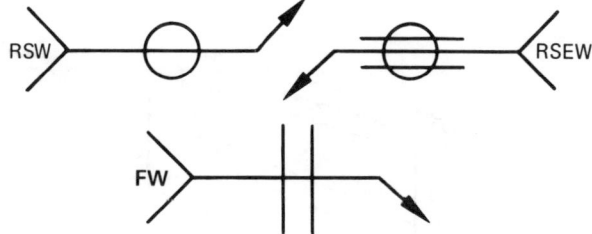

3.2 Location of Weld With Respect to Joint

3.2.1 Arrow Side. Welds on the arrow side of the joint shall be specified by placing the weld symbol below the reference line (see 3.1.1).

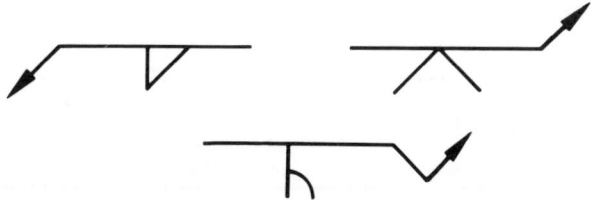

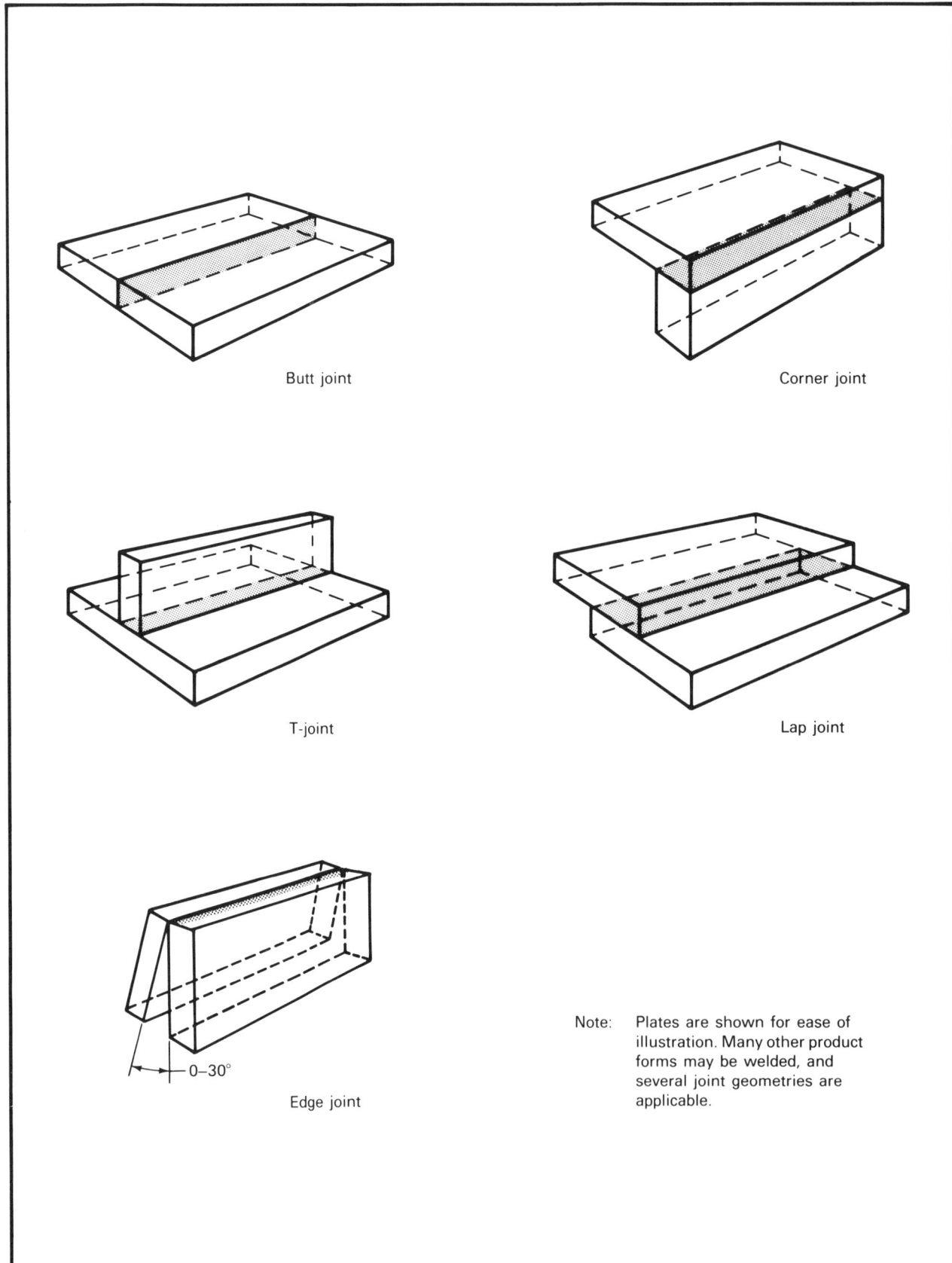

Figure 4 — Basic Joints

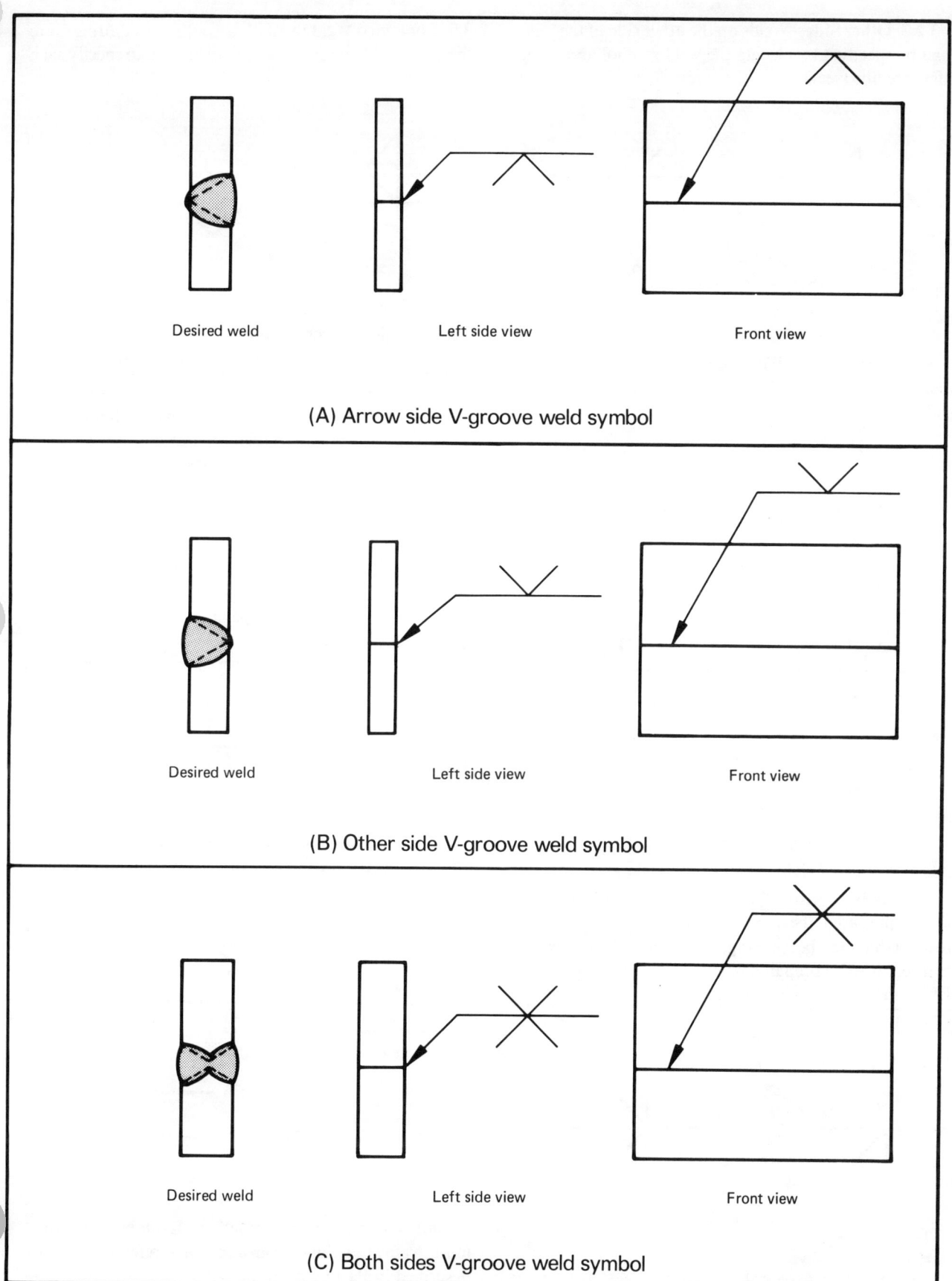

Figure 5 — Application of Arrow Side and Other Side Convention

3.2.2 Other Side. Welds on the other side of the joint shall be specified by placing the weld symbol above the reference line (see 3.1.1).

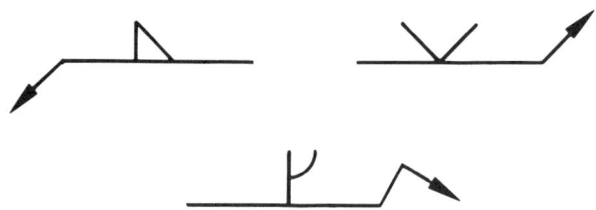

3.2.3 Both Sides. Welds on both sides of the joint shall be specified by placing weld symbols both below and above the reference line (see 3.1.1).

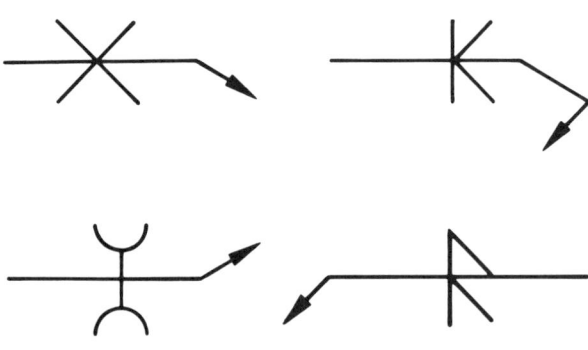

3.3 Orientation of Specific Weld Symbols. Fillet, bevel-groove, J-groove, flare-bevel-groove, and corner-flange weld symbols shall be drawn with the perpendicular leg always to the left.

3.4 Break in Arrow. When only one member of a joint is to be prepared, the arrow shall have a break, and point toward that member (see Figure 6). If it is obvious which member is to be prepared or there is no preference as to which member is to be prepared, the arrow need not be broken.

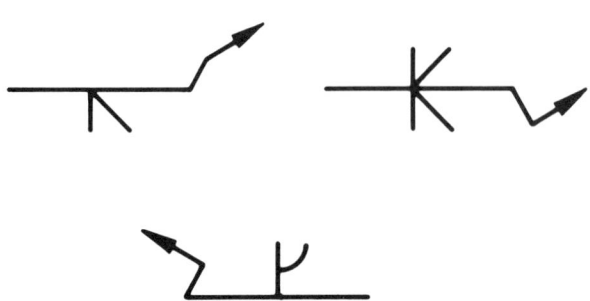

3.5 Combined Weld Symbols. For joints requiring more than one weld type, a symbol shall be used to specify each weld (see Figure 7).

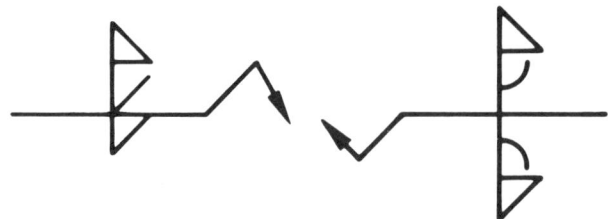

3.6 Multiple Reference Lines

3.6.1 Sequence of Operations. Two or more reference lines may be used to indicate a sequence of operations. The first operation is specified on the reference line nearest the arrow. Subsequent operations are specified sequentially on other reference lines.

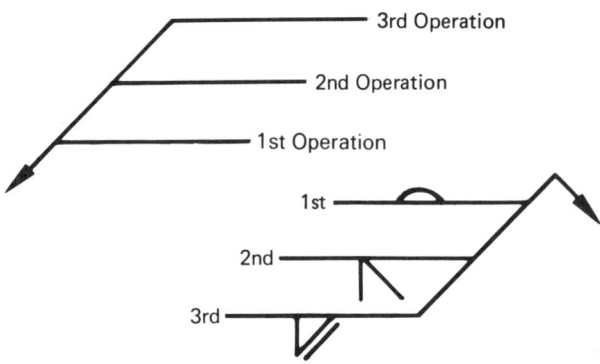

3.6.2 Supplementary Data. The tail of additional reference lines may be used to specify data supplementary to welding symbol information.

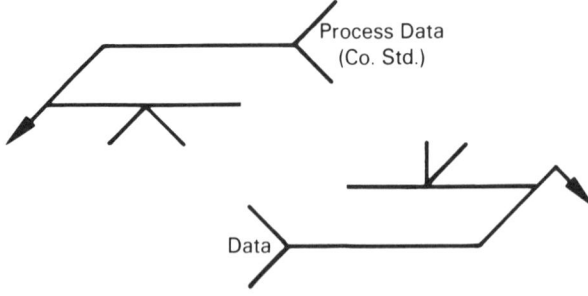

3.6.3 Examination Information. Examination information may be shown on an additional reference line, as described in Part C, or in the tail.

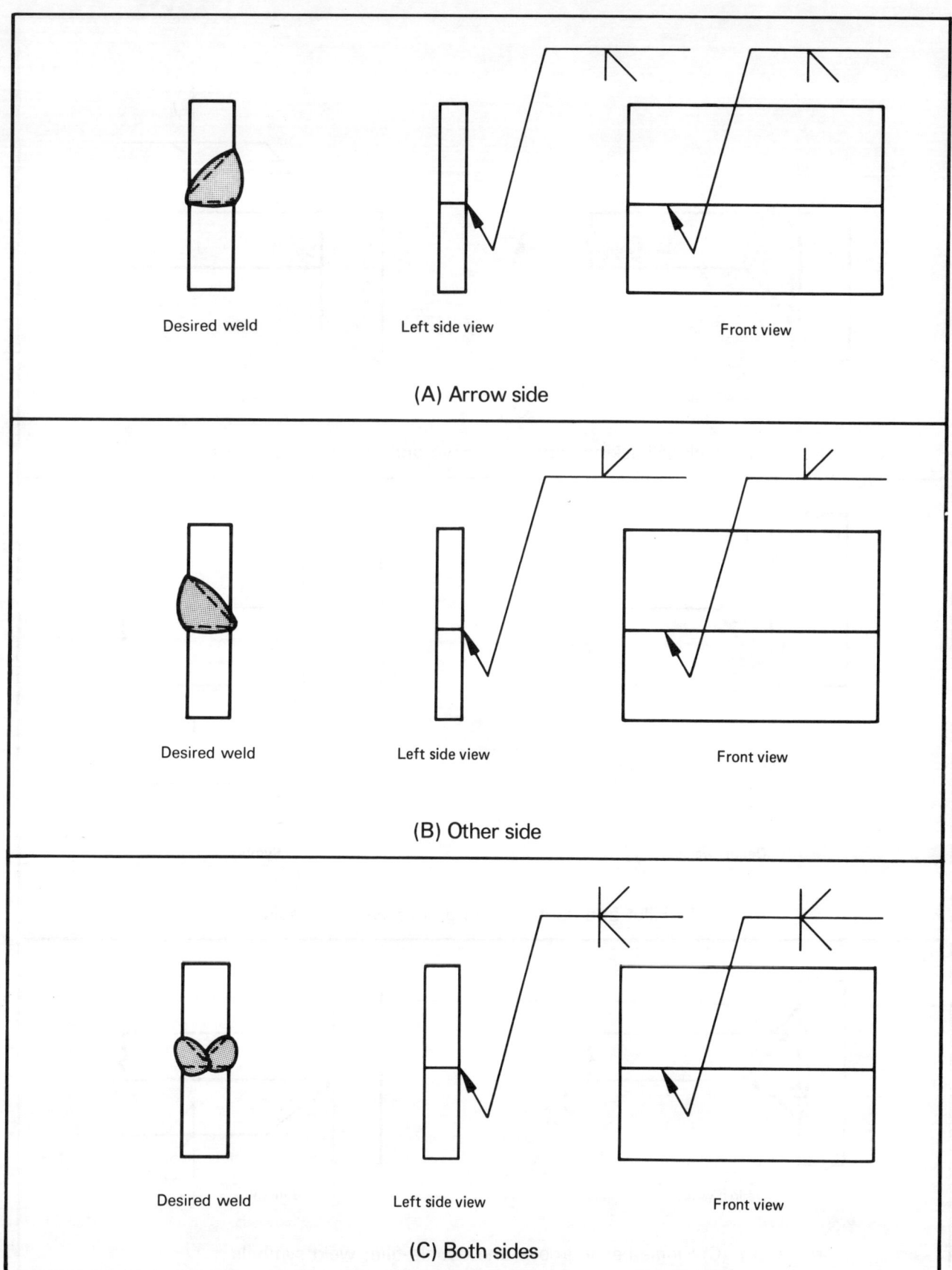

Figure 6 — Application of Break in Arrow of Welding Symbol

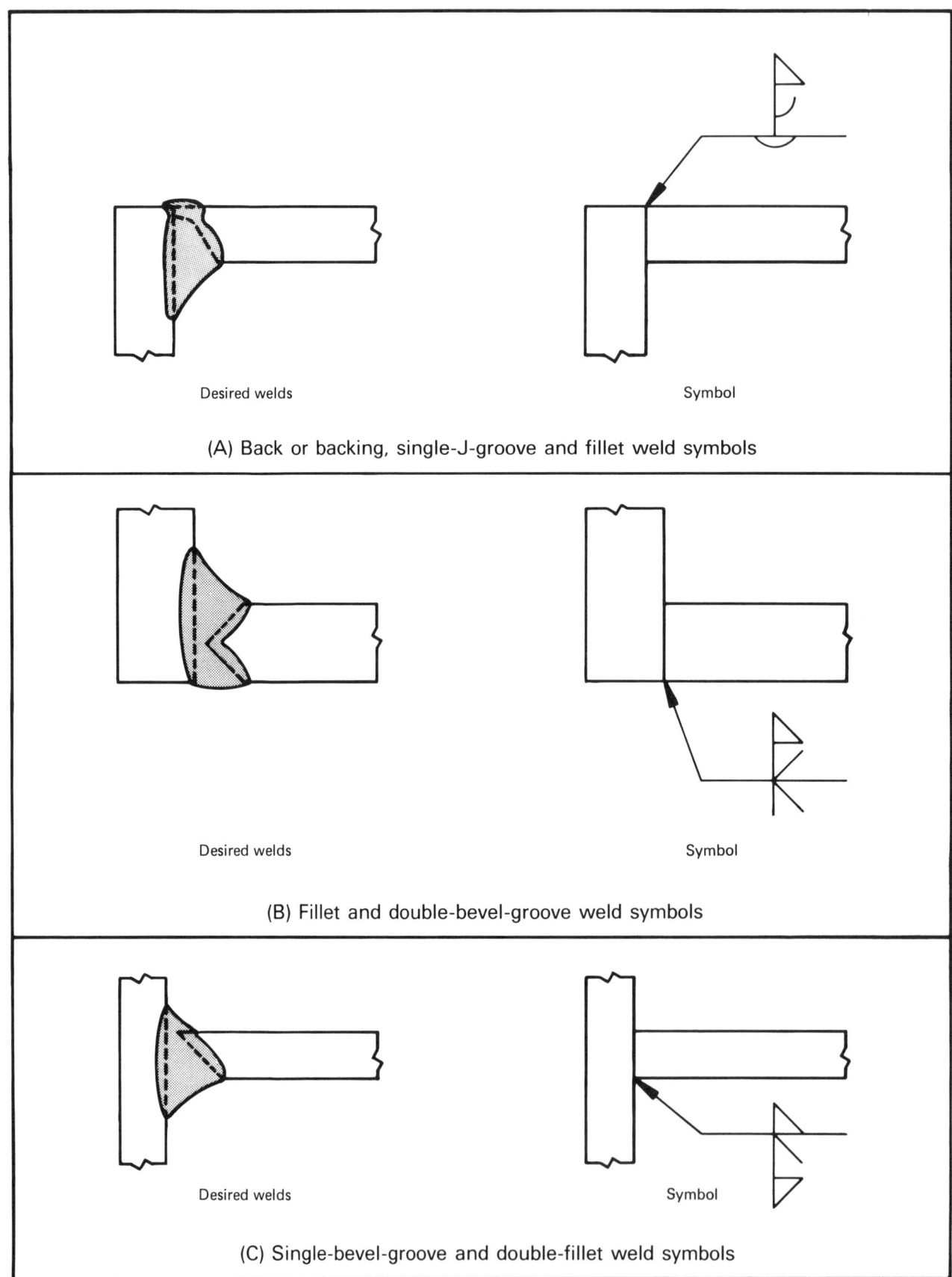

Figure 7 — Combination of Weld Symbols

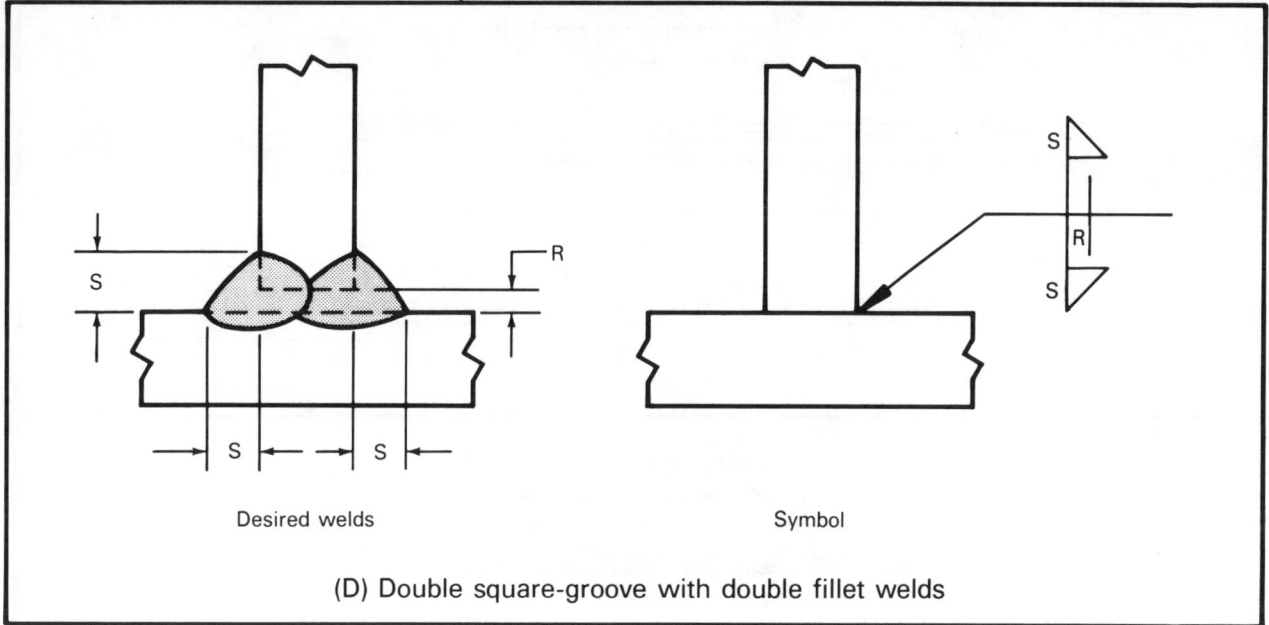

(D) Double square-groove with double fillet welds

Figure 7 (continued) — Combination of Weld Symbols

3.6.4 Field Weld and Weld All-Around Symbols. When required, the weld- (or examine-) all-around symbol shall be placed at the junction of the arrow line and reference line for each operation to which it is applicable. The field weld symbol may also be applied to the same location.

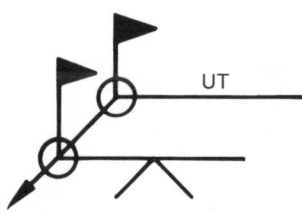

3.7 Field Weld Symbol. Field welds (welds not made in a shop or at the place of initial construction) shall be specified by adding the field weld symbol. The flag shall be placed above and at a right angle to the reference line at the junction with the arrow.

3.8 Extent of Welding Denoted by Symbols

3.8.1 Weld Continuity. Unless otherwise indicated, all welds shall be continuous and of user's standard proportions.

3.8.2 Changes in the Direction of Welding. Symbols only apply between any changes in the direction of welding, or to the extent of hatching or dimension lines (see Figure 8), except when the weld-all-around symbol is used [see Figure 9(B), (C), (D), and (E)]. Additional welding symbols or multiple arrows shall be used to specify the welds required for any changes in direction. When it is desirable to use multiple arrows on a welding symbol, the arrows shall originate from a single reference line [see Figure 9(A)] or from the first reference line in the case of a multiple reference line symbol.

3.8.3. Hidden Members. When the welding of a hidden member is the same as that of a visible member, it may be specified as shown below. If the welding of a hidden member is different from that of a visible member, specific information for the welding of both shall be specified. If needed for clarification, auxilliary illustrations or views shall be provided.

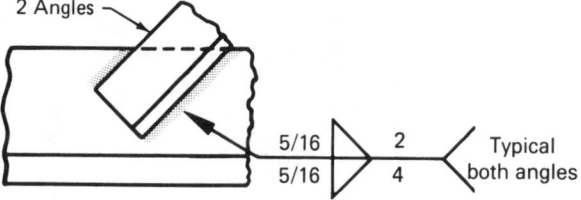

3.8.4 Weld Location Not Specified. A weld, with a length less than the available joint length and not critical

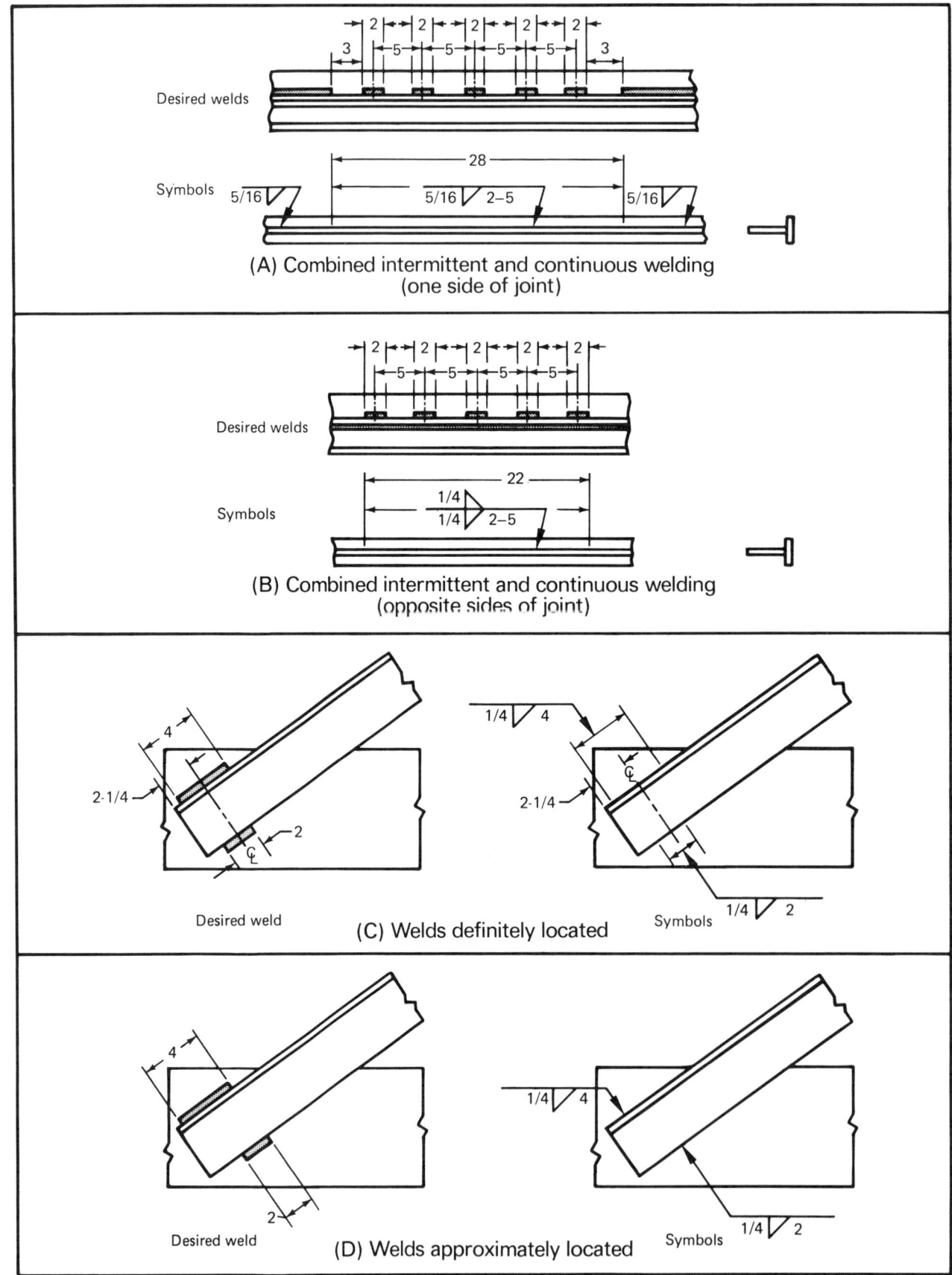

Figure 8 — Designation of Location and Extent of Fillet Welds

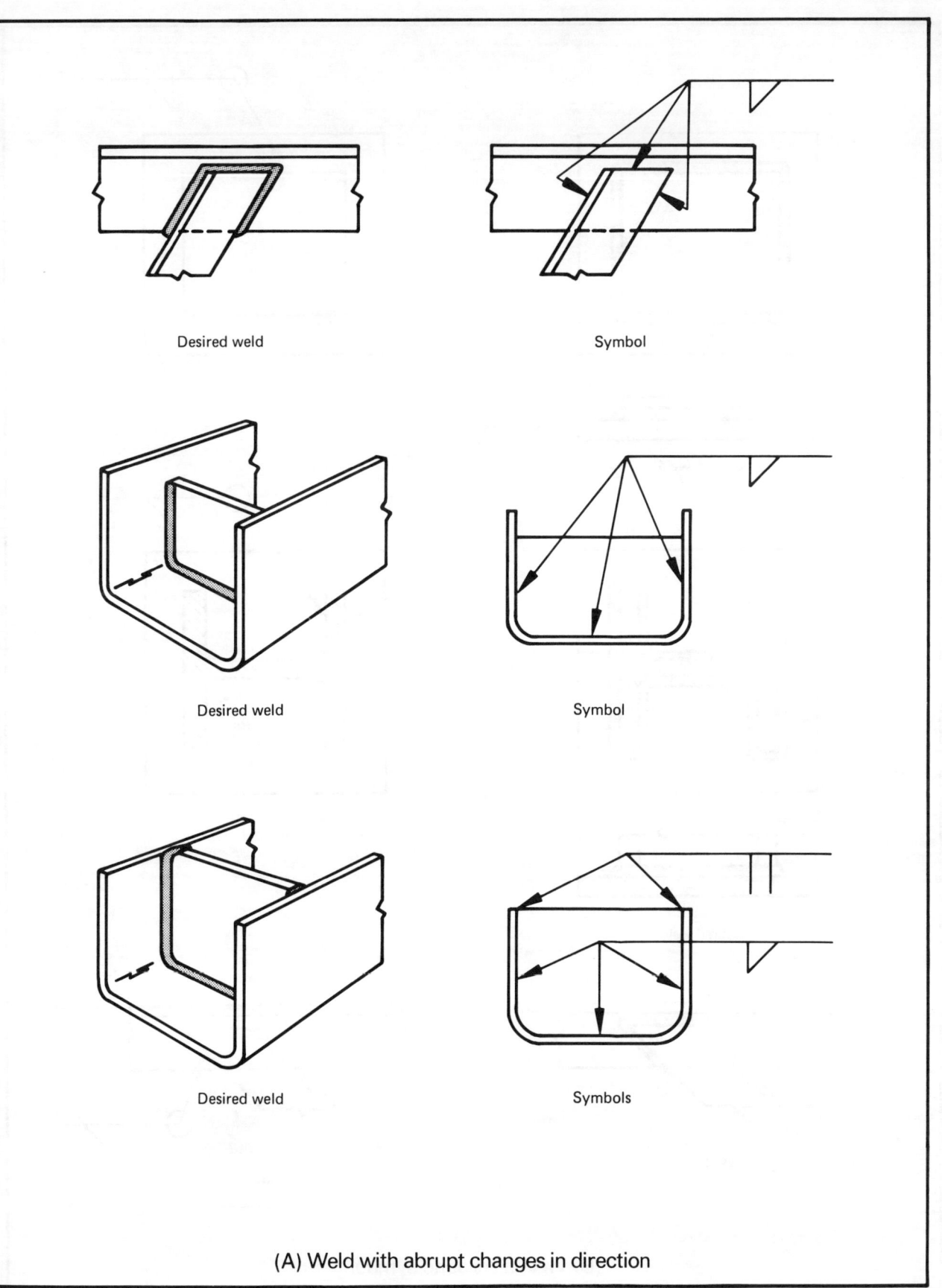

(A) Weld with abrupt changes in direction

Figure 9 — Designation of Extent of Welding

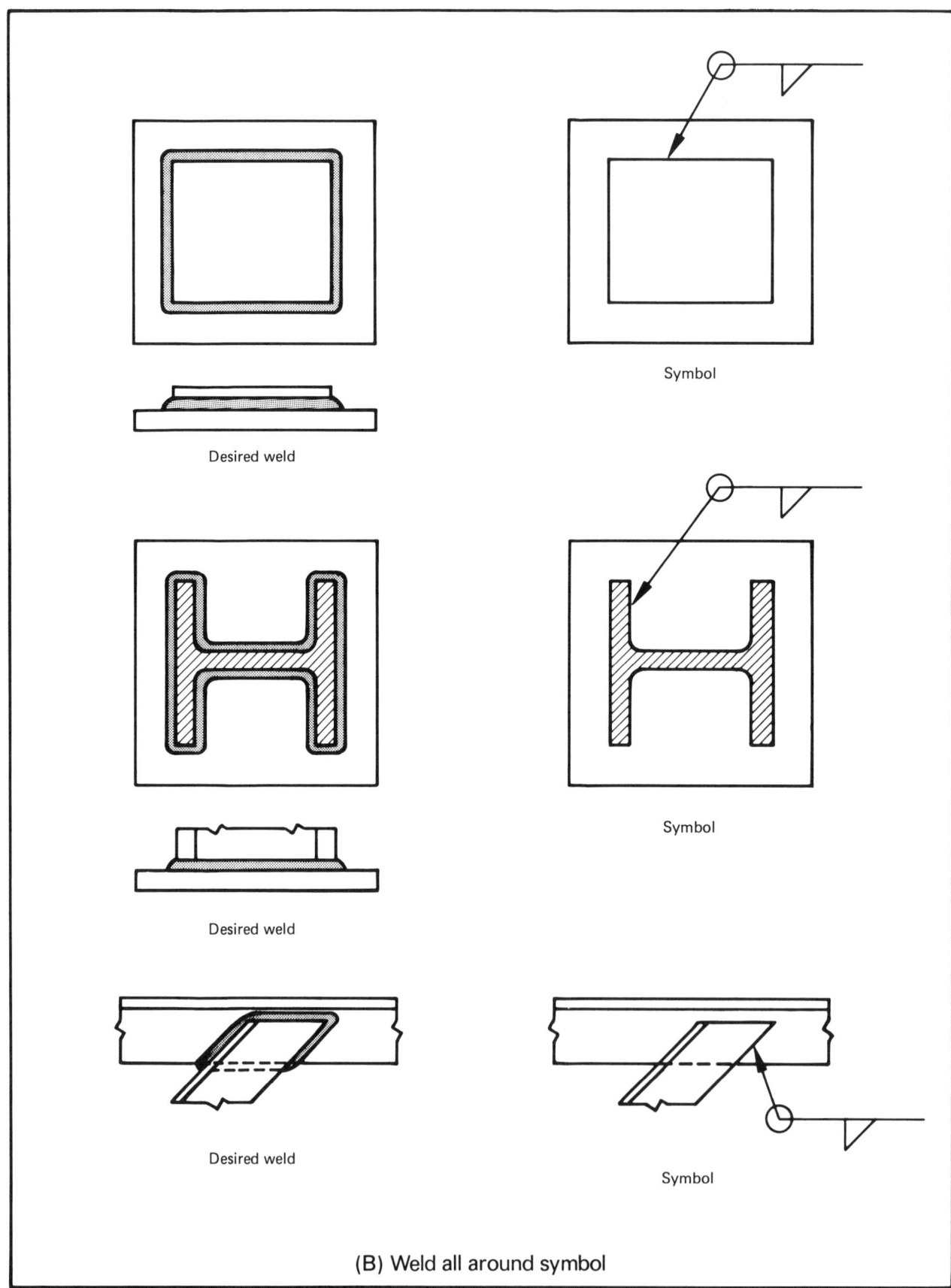

Figure 9 (continued) — Designation of Extent of Welding

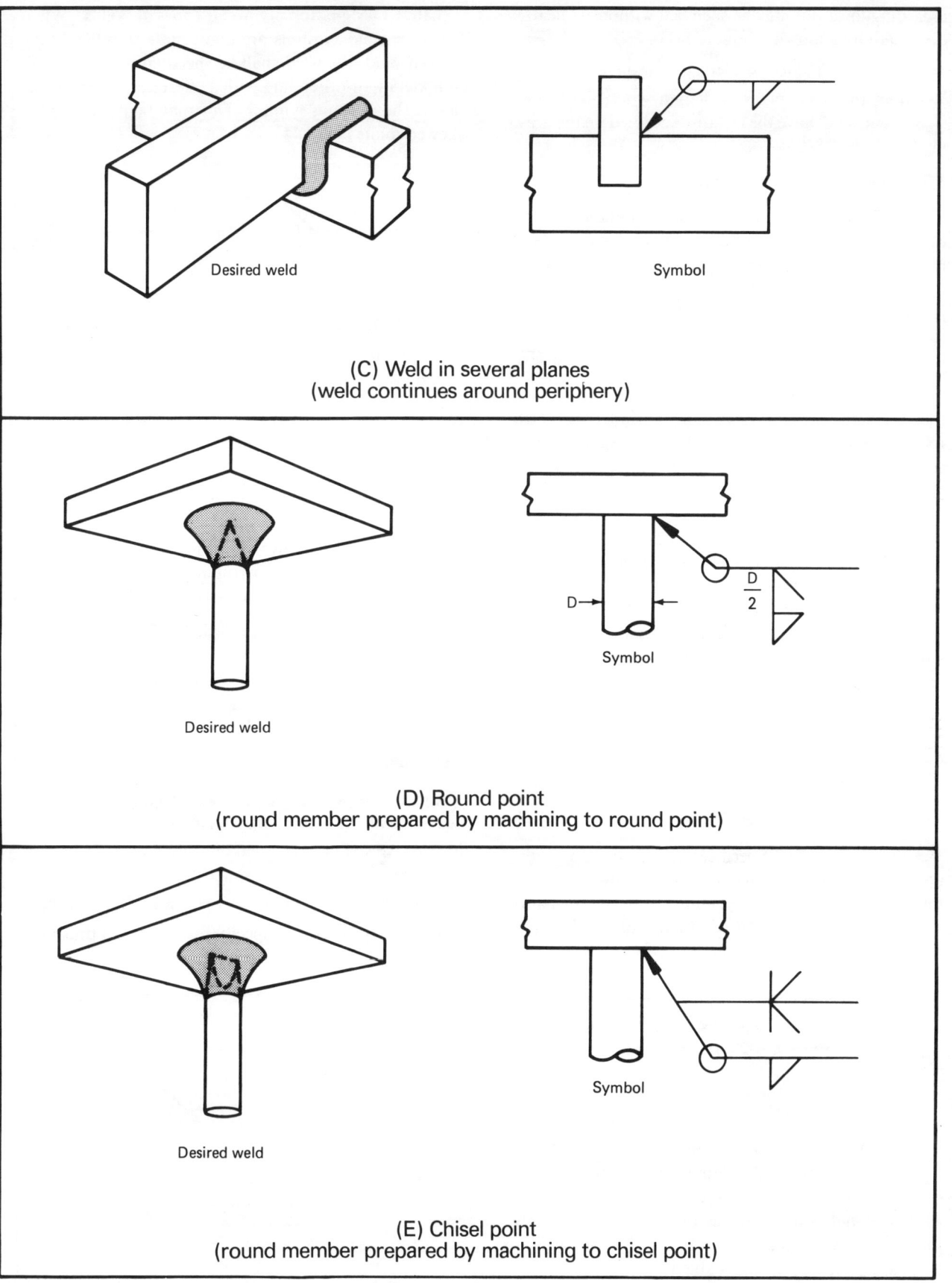

Figure 9 (continued) — Designation of Extent of Welding

regarding location, may be specified without indicating the location as shown in Figure 8(D).

3.8.5 Weld Location Specified. A weld, with a length less than the available joint length whose location is significant, shall have the location specified on the drawing [see Figure 8(C)].

3.9 Weld-All-Around Symbol

3.9.1 Welds in Multiple Directions or Planes. A continuous weld, whether single or combined type, extending around a series of connected joints may be specified by the addition of the weld-all-around symbol at the junction of the arrow and reference line. The series of joints may involve different directions and may lie in more than one plane [see Figure 9(B), (C), (D), and (E)].

3.9.2 Circumferential Welds. Welds extending around the circumference of a pipe are excluded from the requirement regarding changes in direction and do not require the weld-all-around symbol to specify a continuous weld.

3.10 Tail of the Welding Symbol

3.10.1 Welding and Allied Process Specification. The welding and allied process to be used may be specified by placing the appropriate letter designations from Table 1 or Table 2 in the tail of the welding symbol. An auxiliary suffix from Table 3 may be used. (Tables are at the end of text.)

3.10.2 References. Specifications, Codes or any other applicable documents may be specified by placing the reference in the tail of the welding symbol. Information contained in the referenced document need not be repeated in the welding symbol.

3.10.3 Welding Symbols Designated "TYPICAL". Repetitions of identical welding symbols on a drawing may be avoided by designating a single welding symbol as typical and pointing the arrow to the representative joint. The user shall provide additional information to completely identify all applicable joints.

3.10.4 Designation of Special Types of Welds. When the basic weld symbols are inadequate to indicate the desired weld, the weld shall be specified by a cross section, detail, or other data with a reference thereto in the tail of the welding symbol. This may be necessary for skewed joints (see 4.11 and 5.7).

3.10.5 Omission of Tail. When no references are required, the tail may be omitted from the welding symbol.

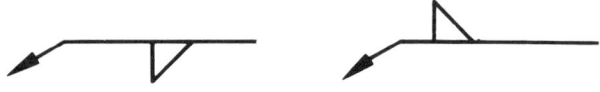

3.10.6 Drawing Notes. Drawing notes may be used to provide information pertaining to the welds. Such information need not be repeated in the welding symbols.

3.11 Contours Obtained by Welding. Welds to be made with approximately flush, flat, convex, or concave contours without the use of mechanical finishing shall be specified by adding the flush, flat, convex, or contour symbol to the weld symbol.

3.12 Finishing of Welds

3.12.1 Contours Obtained by Finishing. Welds to be mechanically finished approximately flush, flat, convex, or concave shall be specified by adding both the appropriate contour and the finishing symbol.

3.12.2 Finishing Methods. The following finishing symbols may be used to specify the method of finishing, but not the degree of finish:

Mechanical Methods:

C — Chipping
G — Grinding
H — Hammering
M — Machining
R — Rolling

3.12.3 Finishing Method Unspecified. Welds to be finished approximately flush, flat, convex, or concave with the method unspecified shall be indicated by adding the letter "U" to the appropriate contour symbol.

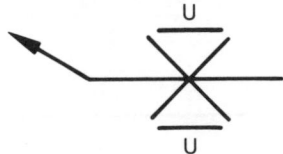

3.13 Melt-Through Symbol. The melt-through symbol shall be used only when complete root penetration plus visible root reinforcement is required in welds made from one side (see Figure 10).

3.13.1 Melt-Through Symbol Location. The melt-through symbol shall be placed on the side of the reference line opposite the weld symbol (see Figure 10).

3.13.2 Melt-Through Dimensions. The height of root reinforcement may be specified by placing the required dimension to the left of the melt-through symbol (see Figure 10). The height of root reinforcement may be unspecified.

3.14 Melt-Through with Flange Welds

3.14.1 Melt-Through With Edge-Flange Welds. Edge-flange welds requiring complete joint penetration shall be specified by the edge-flange weld symbol with the melt-through symbol placed on the opposite side of the reference line. The same welding symbol is used for joints either detailed or not detailed on the drawing (see Figure 10).

3.14.2 Melt-Through With Corner-Flange Welds. Corner-flange welds requiring complete joint penetration shall be specified by the corner-flange weld symbol with the melt-through symbol placed on the opposite side of the reference line. A broken arrow shall point to the member to be flanged where the joint does not give this information (see Figure 10).

3.15 Method of Drawing Symbols. Symbols may be drawn mechanically or freehand. Symbols intended to appear in publications or to be of high precision should be drawn with dimensions and proportions given in Appendix A.

3.16 U.S. Customary and Metric Units. The same system that is the standard for the drawings shall be used on welding symbols. Dual units shall not be used on welding symbols. If it is desired to show conversions from metric to U.S. customary, or vice versa, a table of conversions may be included on the drawing. For guidance in drafting standards, reference is made to the ANSI Y14, *Drafting Manual*. For guidance on the use of metric (SI) units, reference is made to AWS A1.1, *Metric Practice Guide for the Welding Industry*.

4. Groove Welds

4.1 General

4.1.1 Single groove dimensions. Groove weld dimensions shall be specified on the same side of the reference line as the weld symbol [see Figure 11(A) and (F)].

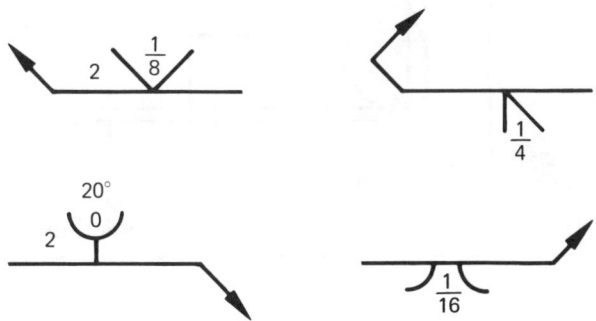

4.1.2 Double-Groove Dimensions. Each groove of a double-groove joint shall be dimensioned; however, the root opening need appear only once (see Figure 12).

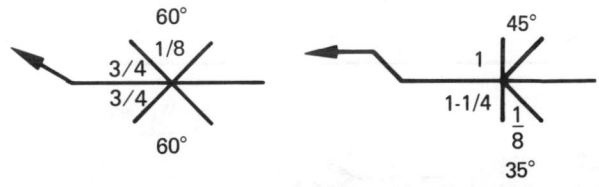

4.1.3 Broken Arrow. For bevel-groove and J-groove welds, a broken arrow is used, when necessary, to identify the member to be prepared (see 3.4).

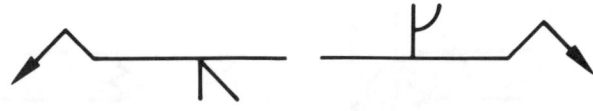

4.2 Depth of Preparation and Groove Weld Size of Groove Welds

4.2.1 Location. The depth of groove preparation, S, and size (E) of a groove weld when specified, shall

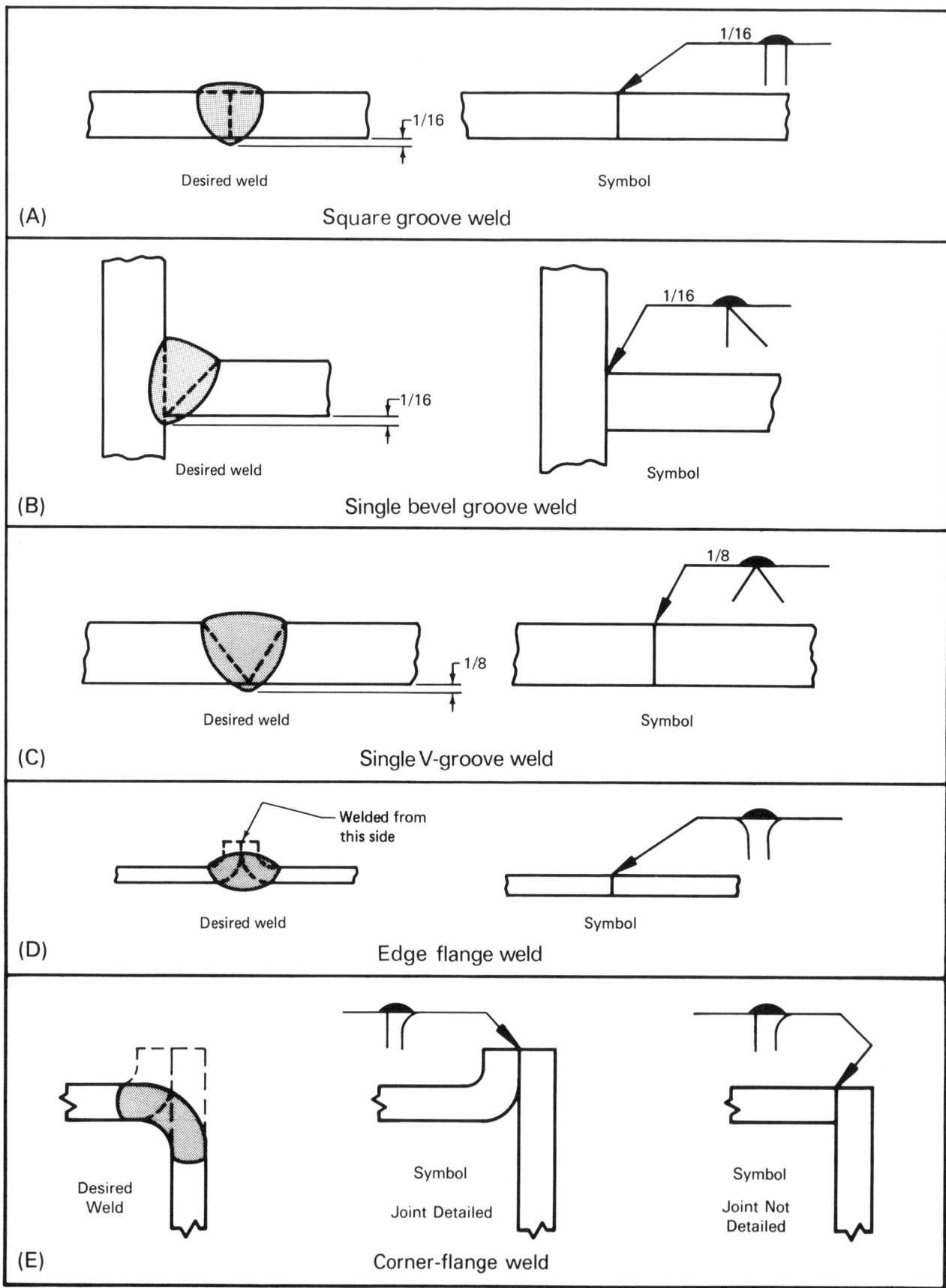

Figure 10 — Applications of Melt-Through Symbol

17

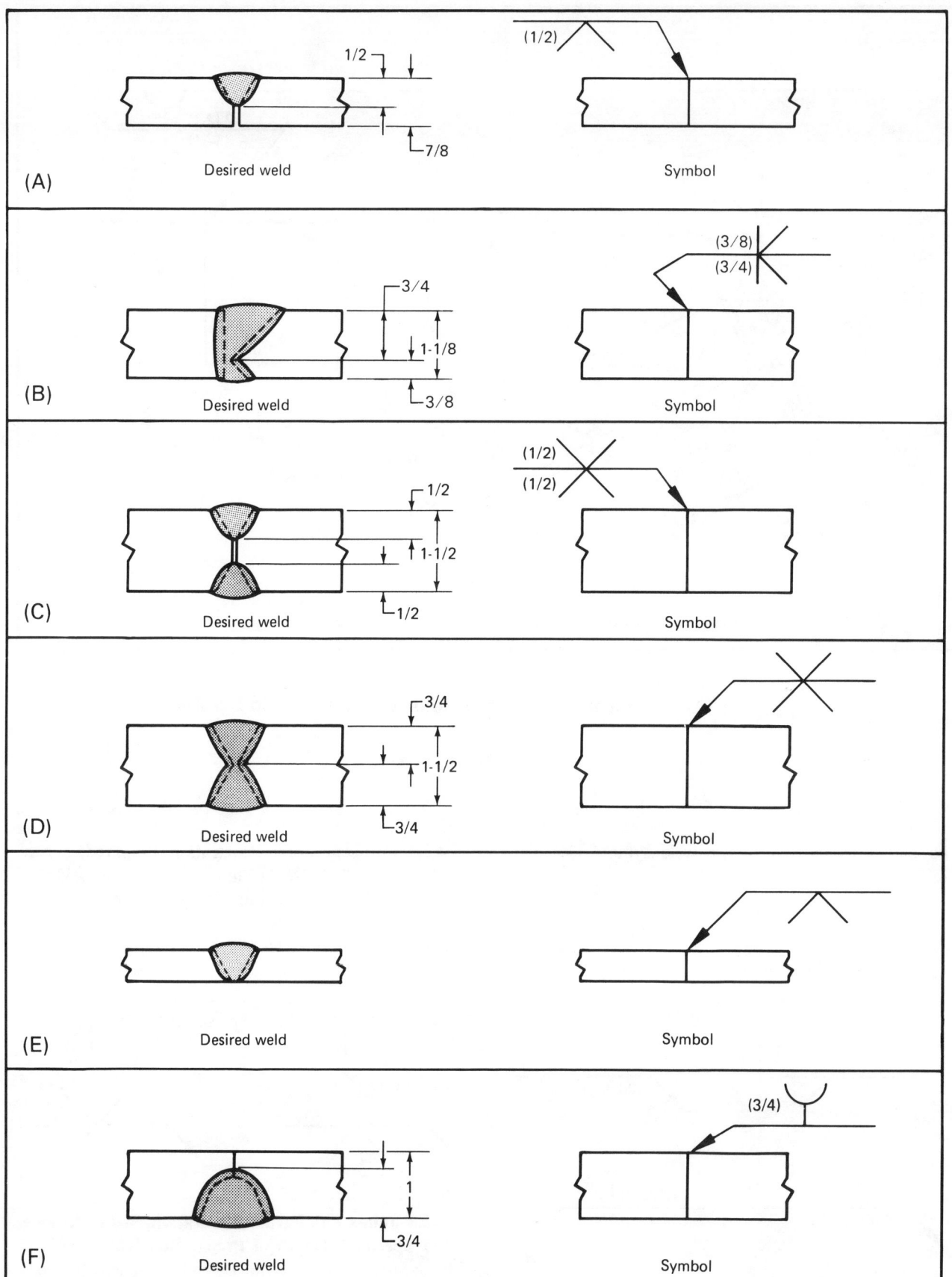

Figure 11 — Designation of Groove Weld Size; Depth of Preparation Not Specified

Groove weld symbol showing use of combined dimensions

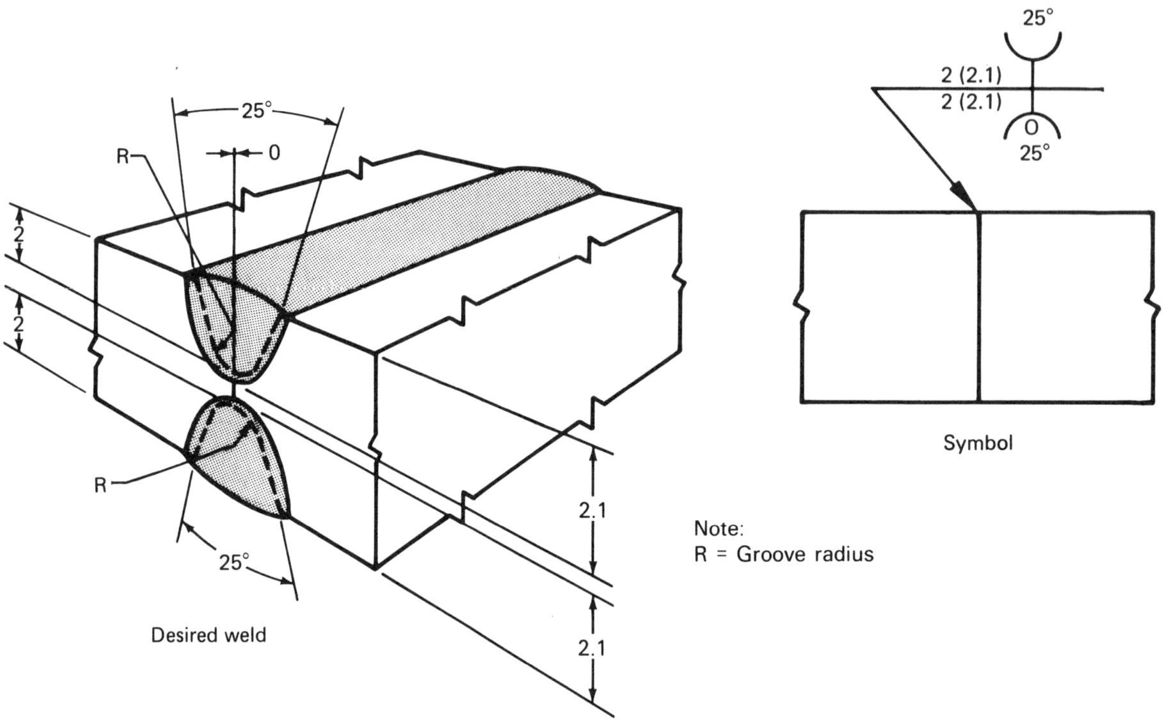

Figure 12 — Application of Dimensions to Groove Weld Symbols

be placed to the left of the weld symbol (see Figures 11–16).

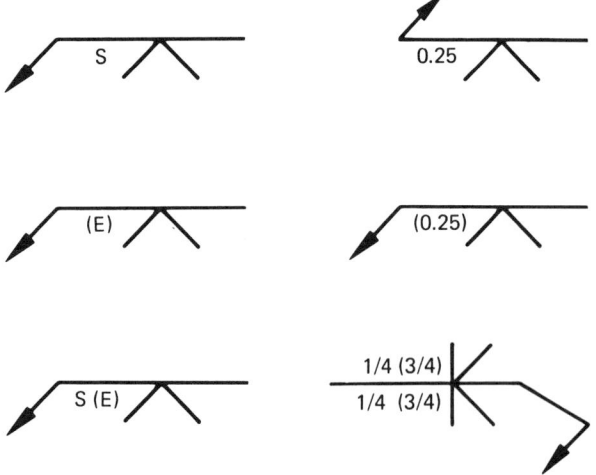

4.2.2 Complete Joint Penetration Required. When no depth of groove preparation and no groove weld size are specified on the welding symbol for single-groove and symmetrical double-groove welds, complete joint penetration is required [see Figure 11(D) and (E)].

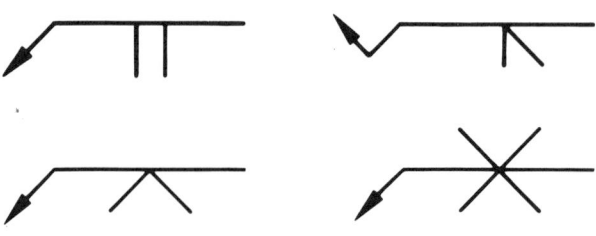

4.2.3 Joints with Partial Penetration, Size of Groove Weld Specified, Depth of Preparation Not Specified. The size of groove welds that extend only partly through the joint shall be specified in parentheses on the welding symbol [see Figure 11(A), (C), and (F)].

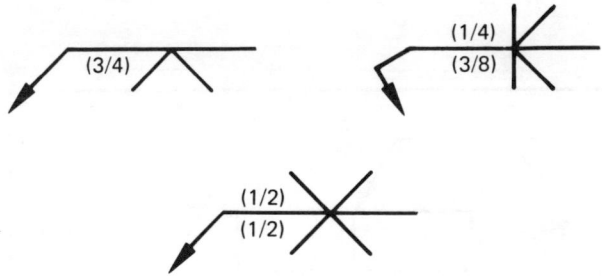

4.2.4 Welds with Complete Joint Penetration, Size of Groove Weld Specified, Depth of Preparation Not Specified. The size of nonsymmetrical groove welds that extend completely through the joint shall be specified in parentheses on the weld symbol (see Figure 15).

4.2.5 Depth of Preparation Specified, Size of Groove Weld Specified Elsewhere. A dimension not in parentheses placed to the left of bevel-, V-, J-, or U-groove weld symbol indicates only the depth of preparation.

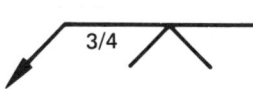

4.2.6 Depth of Preparation and Size of Groove Weld Specified. Except for square-groove welds, the groove weld size "(E)" in relation to the depth of groove preparation "S" is shown as "S(E)" to the left of the weld symbol. "E" only is shown for the square-groove weld (see Figures 13, 14, 16, and 19).

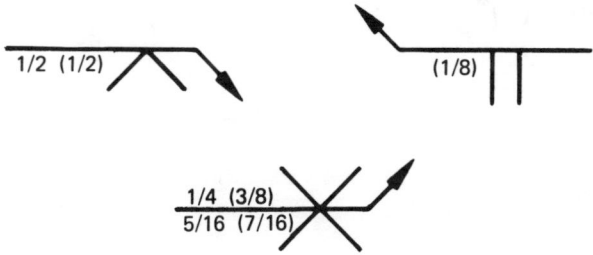

4.2.7 Joint Preparation Not Specified, Complete Joint Penetration Required. Optional joint preparation with complete joint penetration required is specified by placing the letters "CJP" in the tail of the arrow and omitting the weld symbol (see Figure 17).

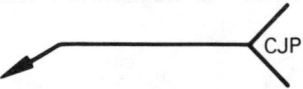

4.2.8 Joint Penetration Not Specified, Size of Groove Weld Specified. For optional joint preparation, the groove weld size may be specified by placing the dimension "(E)" on the arrow side or other side of the reference line as required, but omitting the weld symbol (see Figure 18).

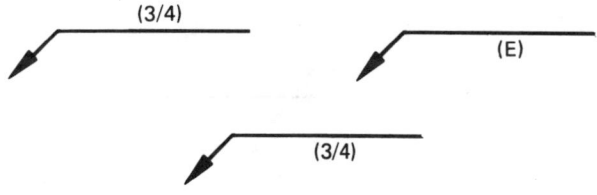

4.2.9 Flare-Groove Welds. The dimension "S" of flare groove welds is considered as extending only to the tangent point indicated below by dimension lines (see Figure 19).

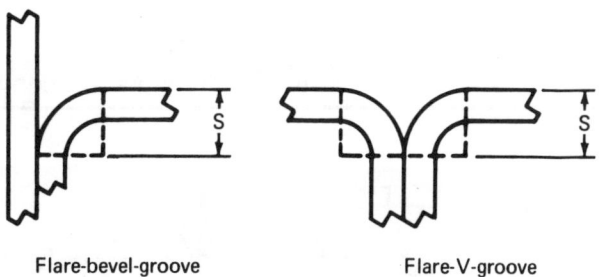

Flare-bevel-groove   Flare-V-groove

4.3 Groove Dimensions

4.3.1 Root Opening. The root opening of groove welds shall be specified inside the weld symbol and only on one side of the reference line (see Figure 20).

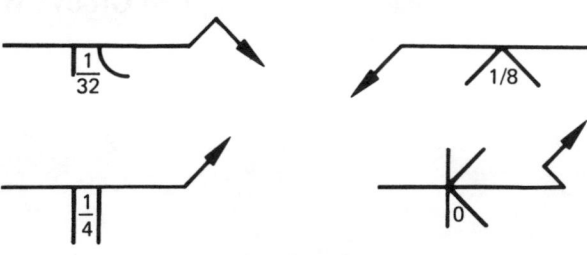

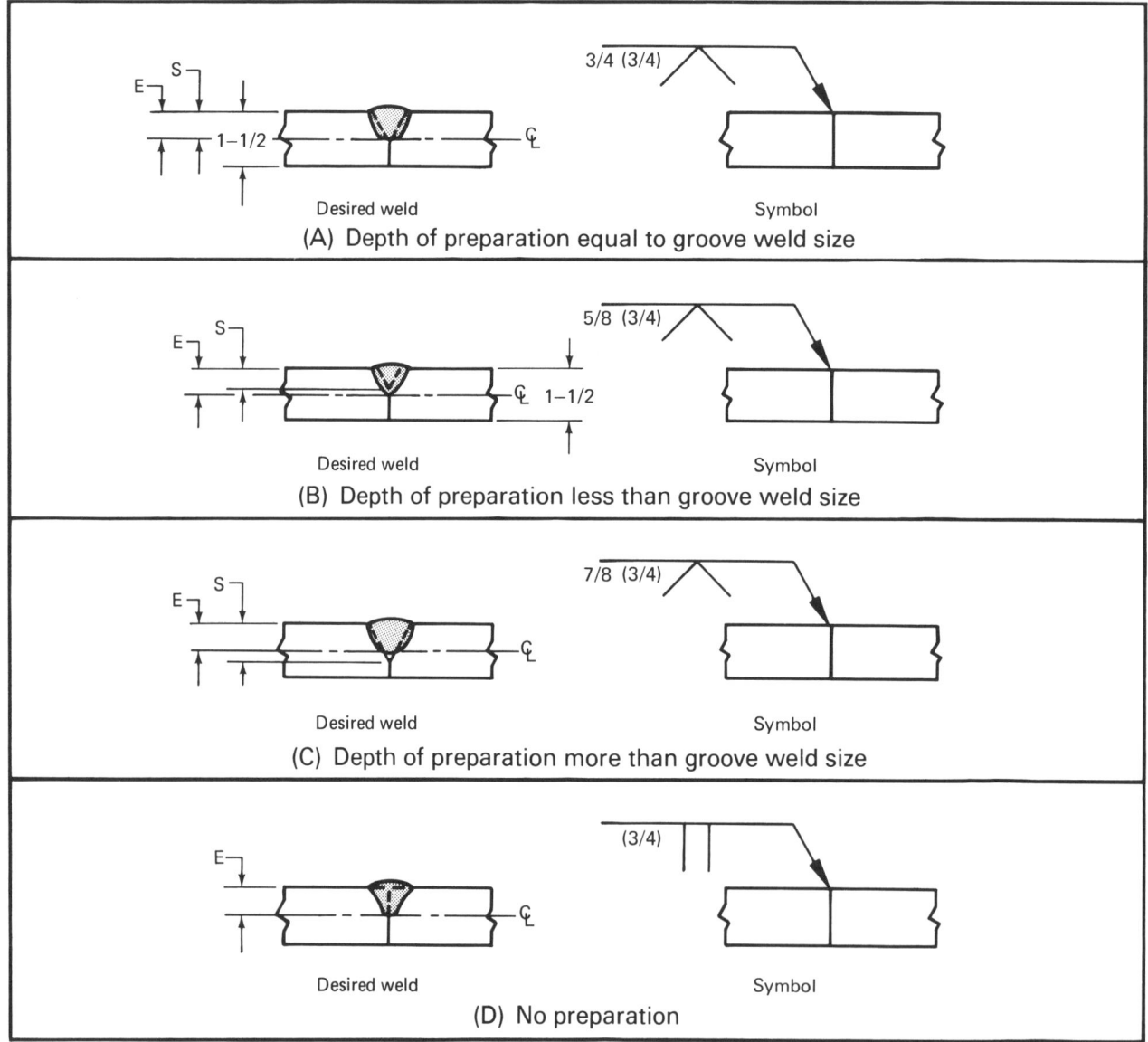

Figure 13 — Examples of Different Relationships Between Depth of Preparation "S" and Groove Weld Size "(E)"

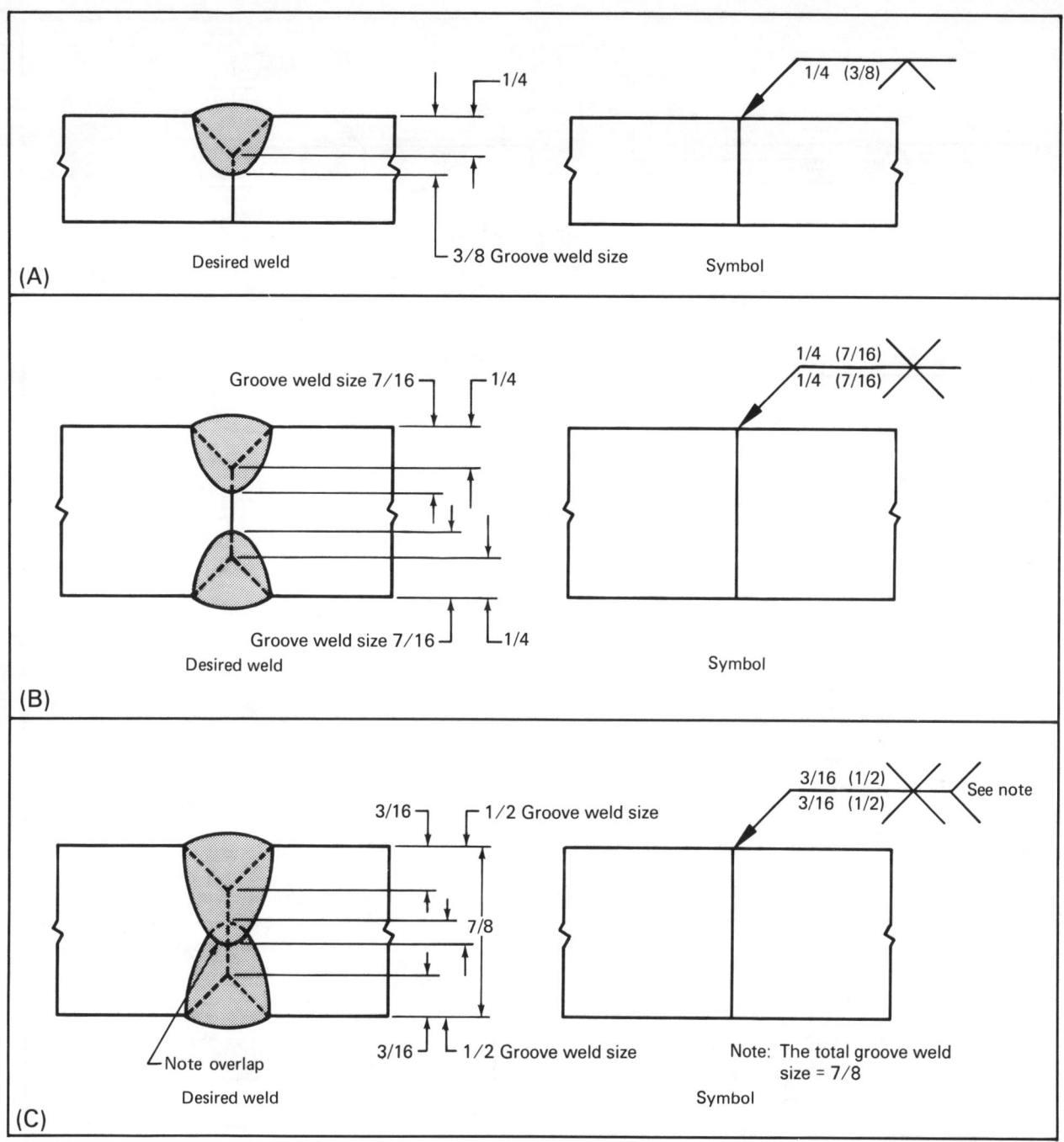

Figure 14 — Designation of Groove Weld Size With Specified Depth of Preparation

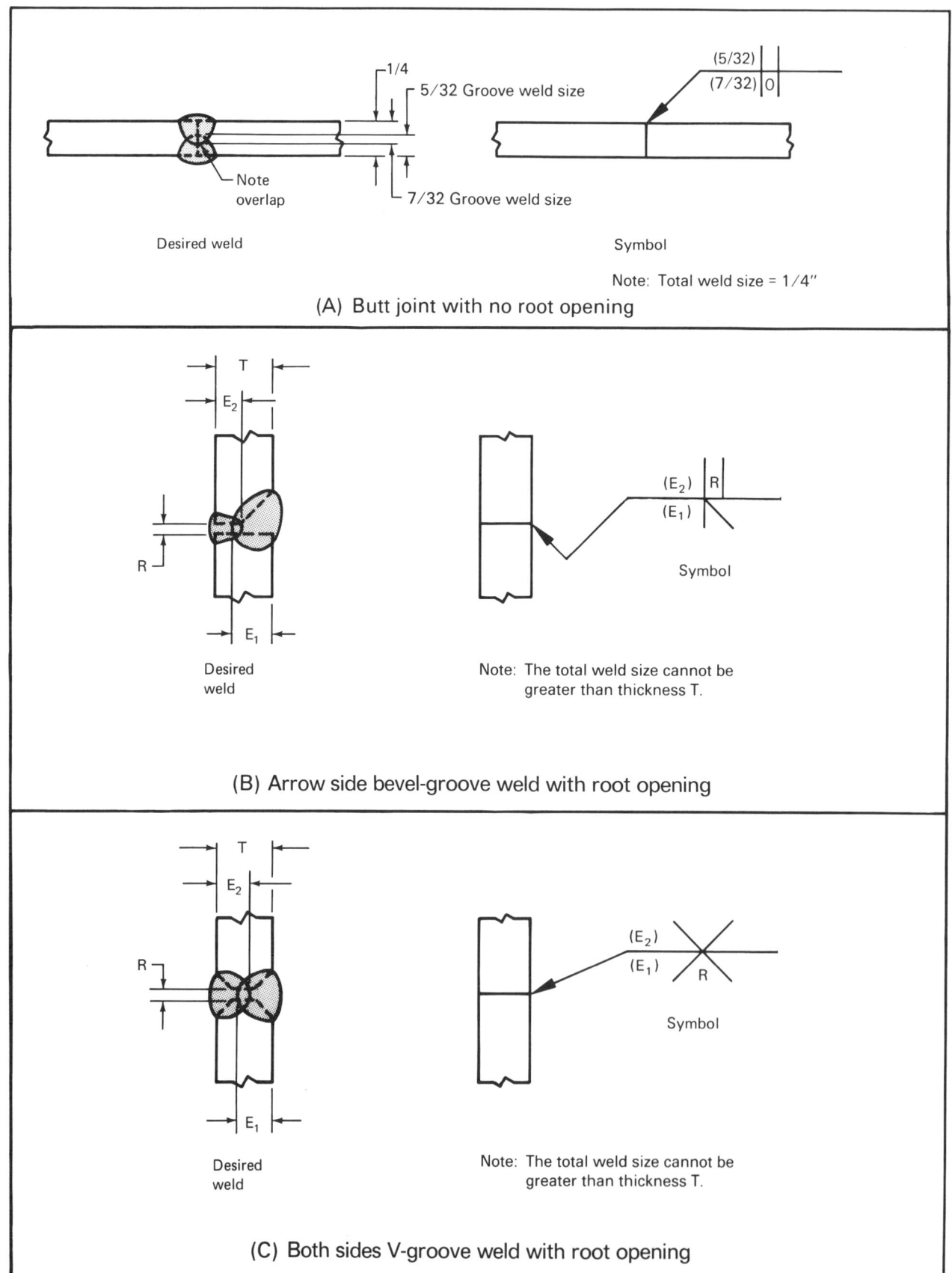

Figure 15 — Designation of Groove Weld Size Without Specified Depth of Preparation

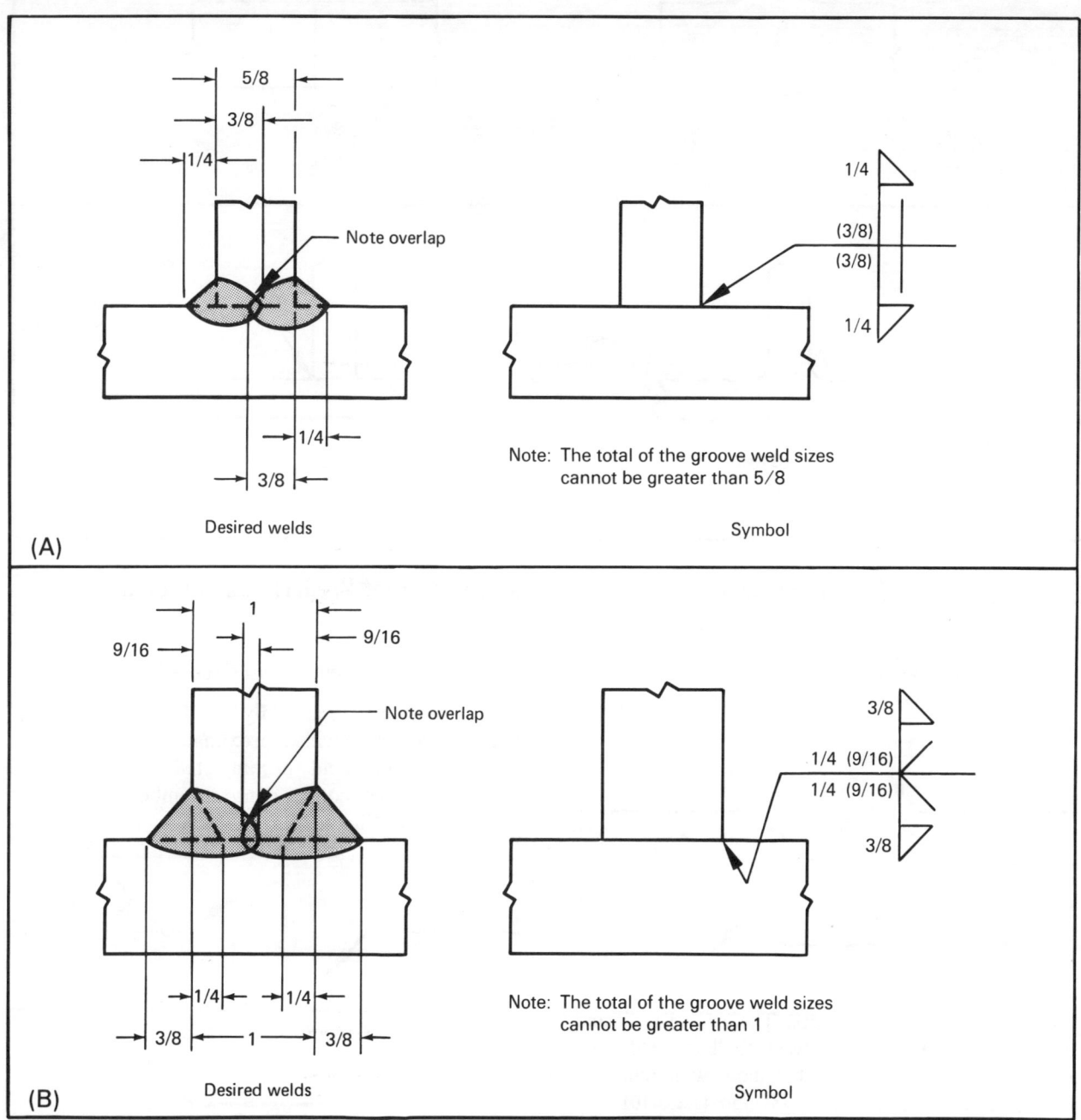

Figure 16 — Combined Groove and Fillet Welds With Specified Groove Weld Size and Fillet Weld Size, and Depth of Preparation

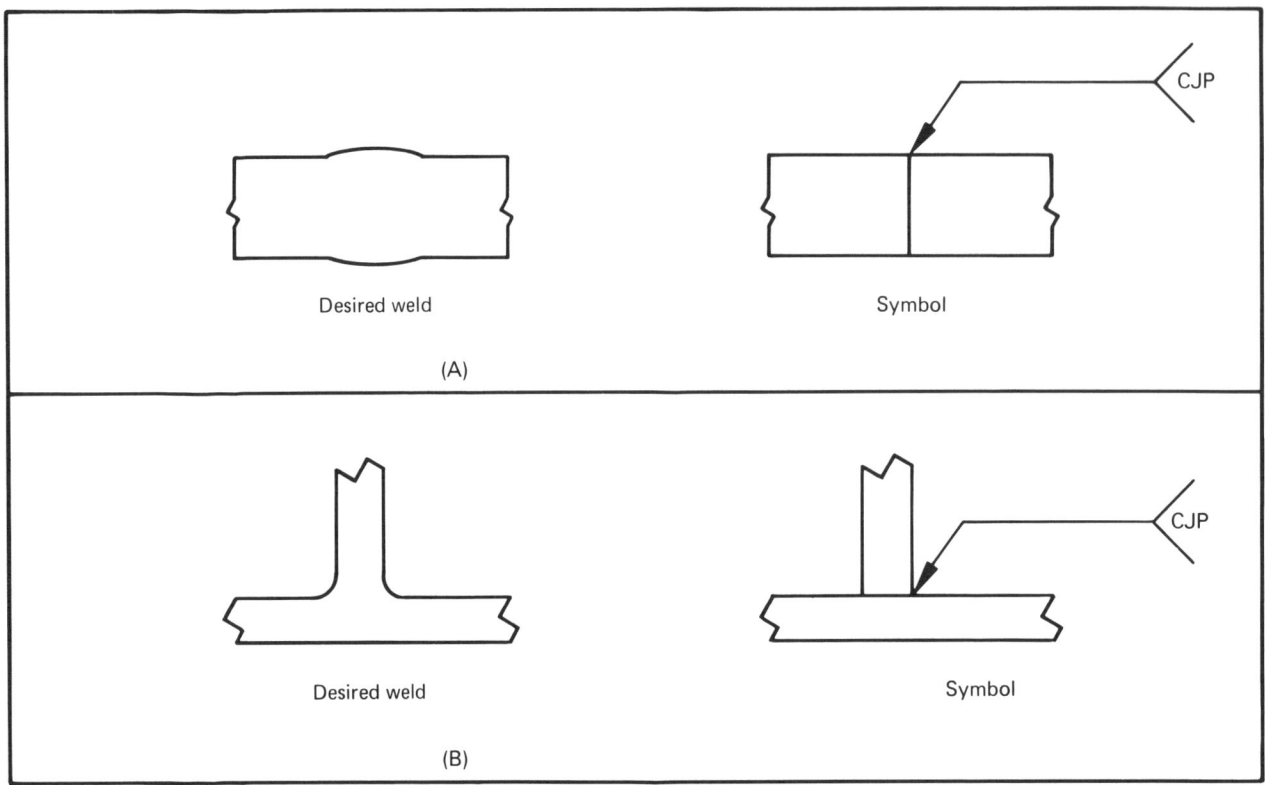

Figure 17 — Complete Joint Penetration Required, Joint Preparation Optional

4.3.2 Groove Angle. The groove angle of groove welds, shall be specified outside the weld symbol (see Figure 21).

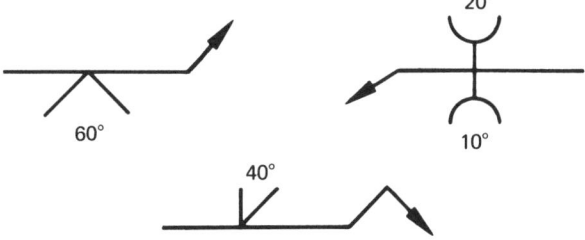

4.3.3 Radii and Root Faces. The groove radii and root faces of U- and J-groove welds shall be specified by a cross section, detail, or other data, with reference thereto in the tail of the welding symbol (see 3.10).

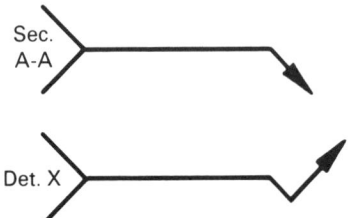

4.4 Contours and Finishing of Groove Welds

4.4.1 Contours Obtained by Welding. Groove welds that are to be welded with approximately flush or convex faces without postweld finishing shall be specified by adding the flush or convex contour symbol to the welding symbol (see Figure 22).

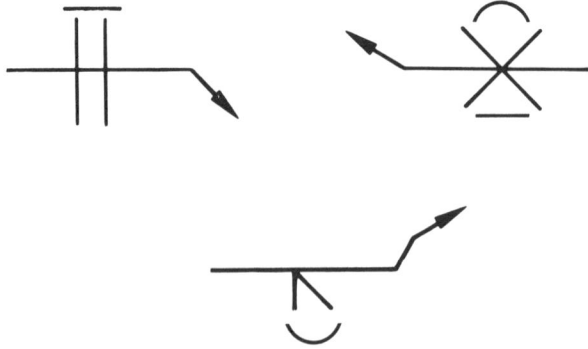

4.4.2 Contours Obtained by Postweld Finishing. Groove welds whose faces are to be finished flush or convex by postweld finishing shall be specified by adding both the appropriate contour and finishing symbol to the welding symbol. Welds that require a flat but not flush

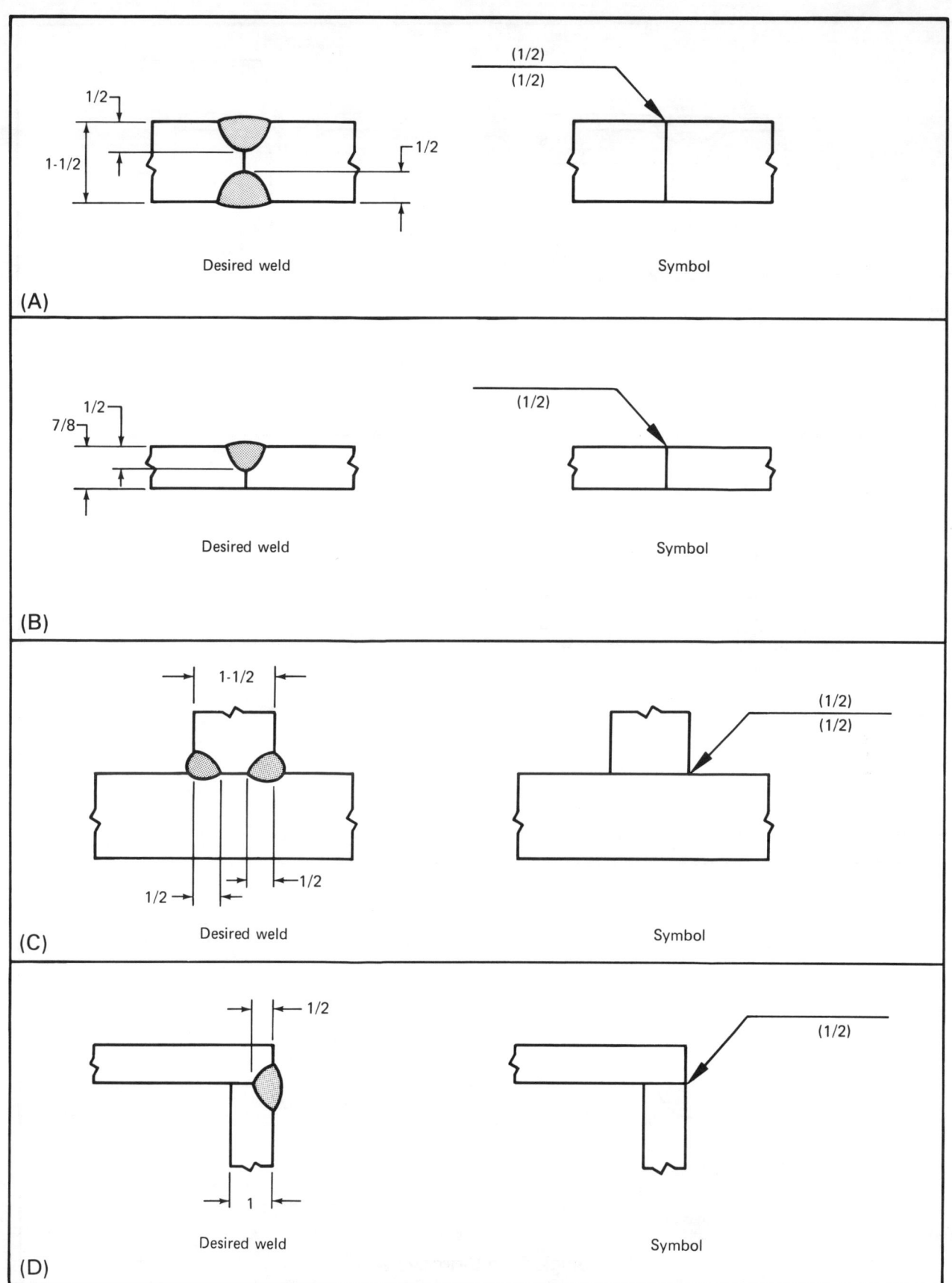

Figure 18 — Partial Joint Penetration Specified, Joint Preparation Optional

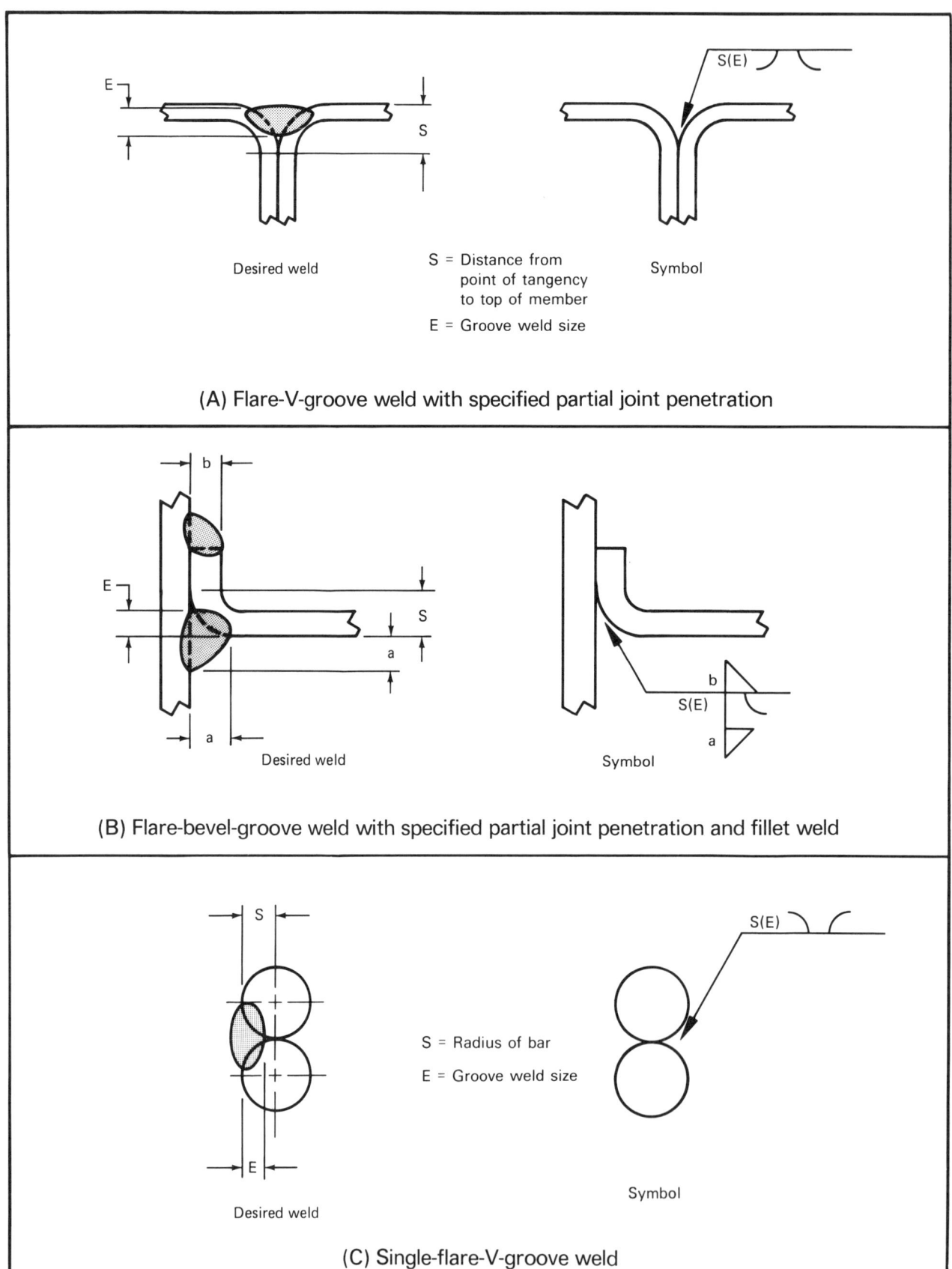

Figure 19 — Application of Flare-Bevel and Flare-V-Groove Weld Symbols

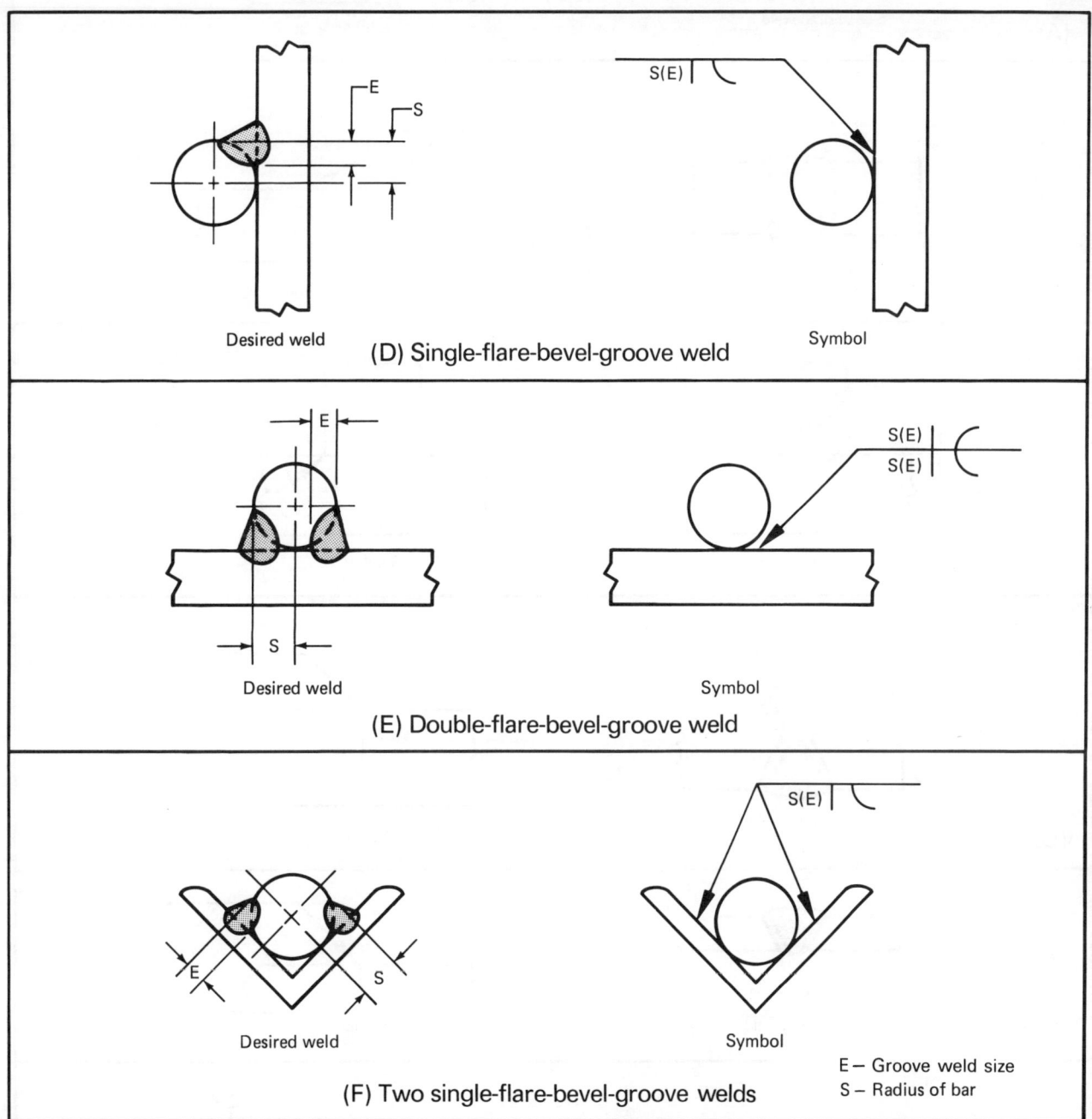

Figure 19 (continued) — Application of Flare-Bevel and Flare-V-Groove Weld Symbols

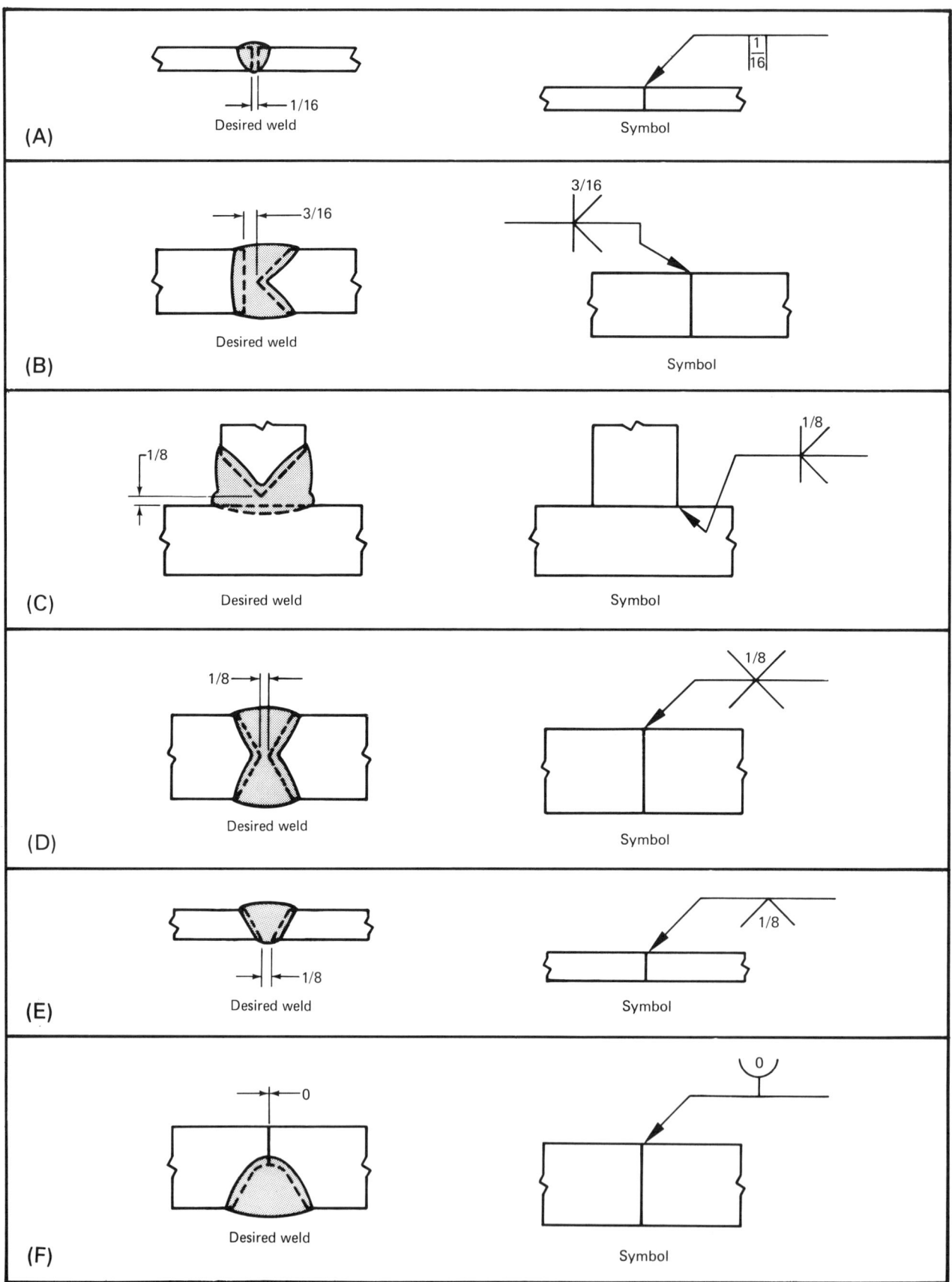

Figure 20 — Designation of Root Opening of Groove Welds

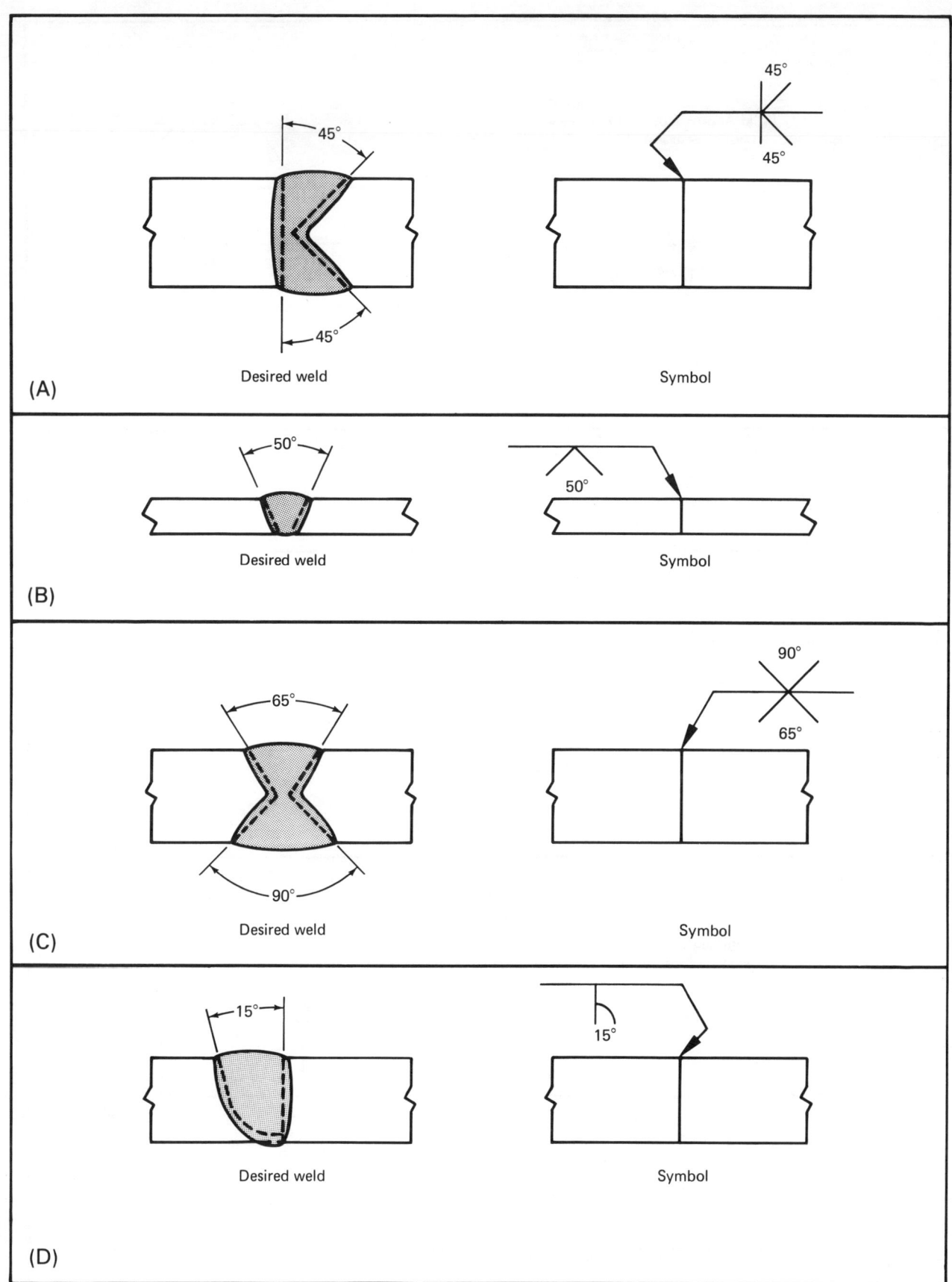

Figure 21 — Designation of Groove Angle of Groove Welds

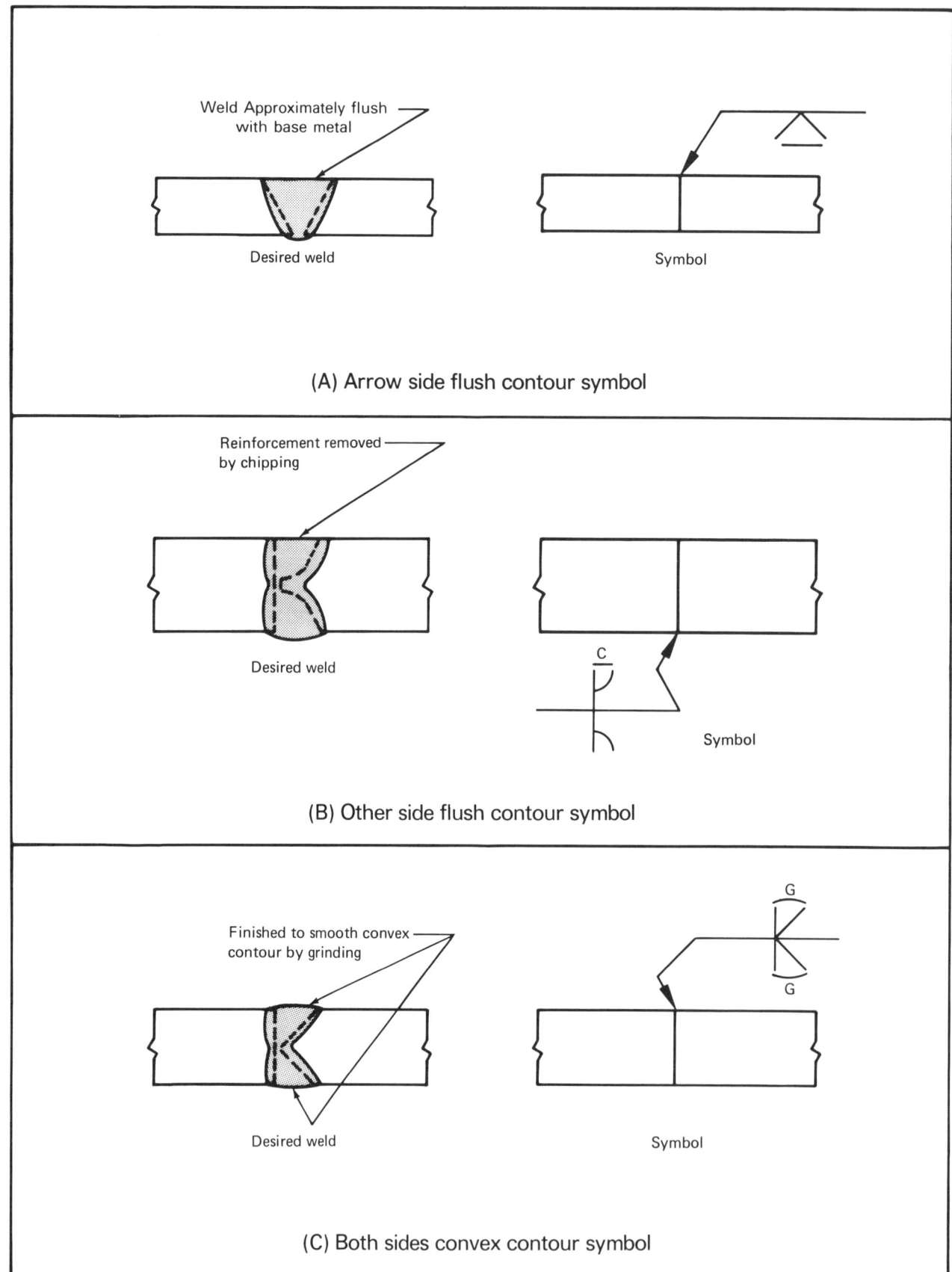

Figure 22 — Application of Flush and Convex Contour Symbols to Groove Weld Symbols

surface, require an explanatory note in the tail of the welding symbol (see 3.12.1).

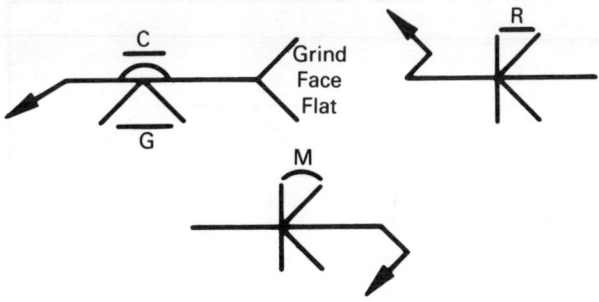

4.5 Back and Backing Welds

4.5.1 General. The back and backing weld symbols are identical. The sequence of welding determines which designation applies. The back weld is made after the groove weld and the backing weld is made before the groove weld (see 4.5.2 and 4.5.3).

4.5.2 Back Weld Symbol. The back weld symbol is placed on the side of the reference line opposite a groove weld symbol. When a single reference line is used, "back weld" shall be specified in the tail of the symbol. Alternately, if a multiple reference line is used, the back weld symbol shall be placed on a reference line subsequent to the reference line specifying the groove weld [see Figure 23(A)].

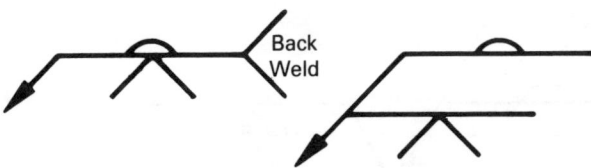

4.5.3 Backing Weld Symbol. The backing weld symbol is placed on the side of the reference line opposite a groove weld symbol. When a single reference line is used, "backing weld" shall be specified in the tail of the arrow. If a multiple reference line is used, the backing weld symbol shall be placed on a reference line prior to that specifying the groove weld [see Figure 23(B) and (C)].

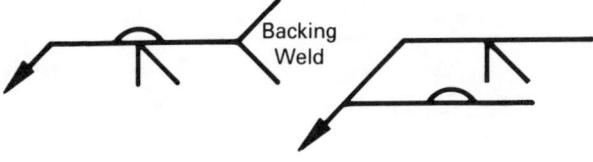

4.5.4 Contour and Finishing of Back or Backing Welds

4.5.4.1 Contours Obtained by Welding. Back or backing welds that are to be welded with approximately flush or convex faces without postweld finishing shall be specified by adding the flush or convex contour symbol to the back or backing weld symbol (see 3.11).

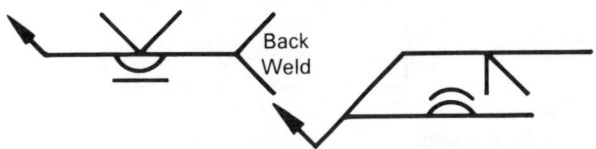

4.5.4.2 Contours Obtained by Postweld Finishing. Back or backing welds that are to be finished approximately flush or convex by postweld finishing shall be specified by adding the appropriate contour and finishing symbols to the back or backing weld symbol (see 3.12.1). Welds that require a flat but not flush surface, require an explanatory note in the tail of the symbol.

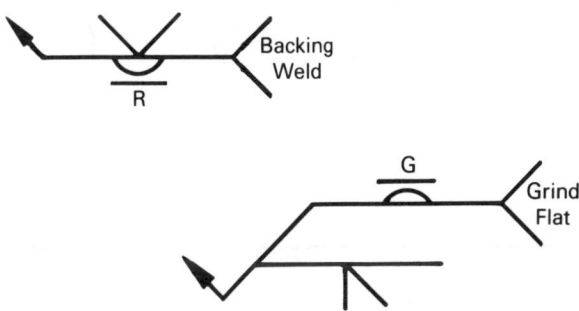

4.6 Joint with Backing. A joint with backing is specified by placing the backing symbol on the side of the reference line opposite the groove weld symbol. If the backing is to be removed after welding, an "R" shall be placed in the backing symbol [see Figure 24(A)]. Material and dimensions of backing shall be specified in the tail of the symbol or on the drawing.

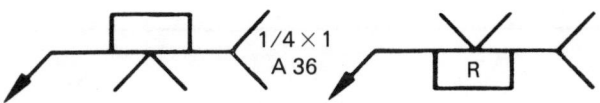

4.7 Joint with Spacer. A joint with a required spacer is specified with the groove weld symbol modified to show a rectangle within it. In case of multiple reference lines,

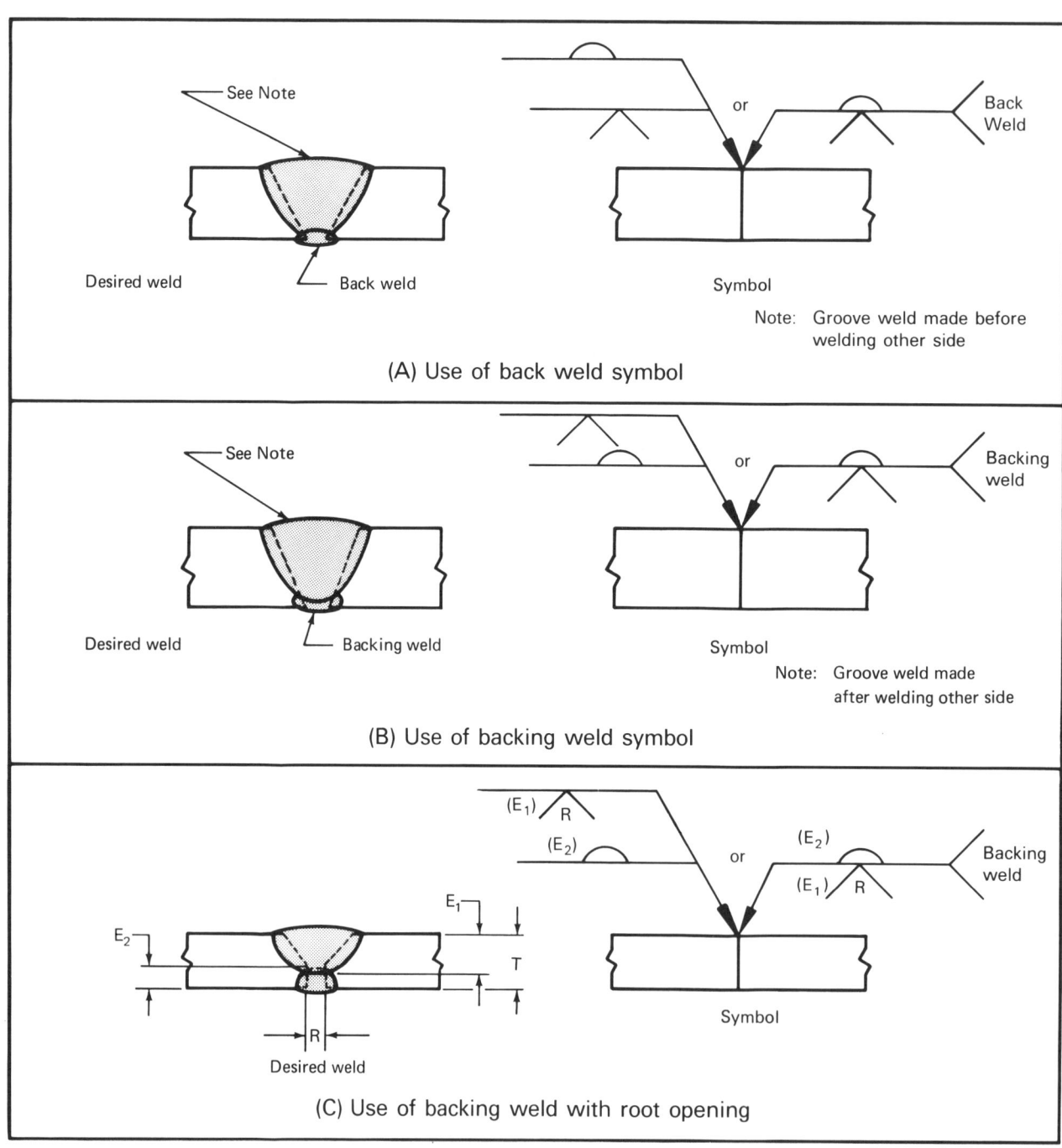

Figure 23 — **Application of Back or Backing Weld Symbol**

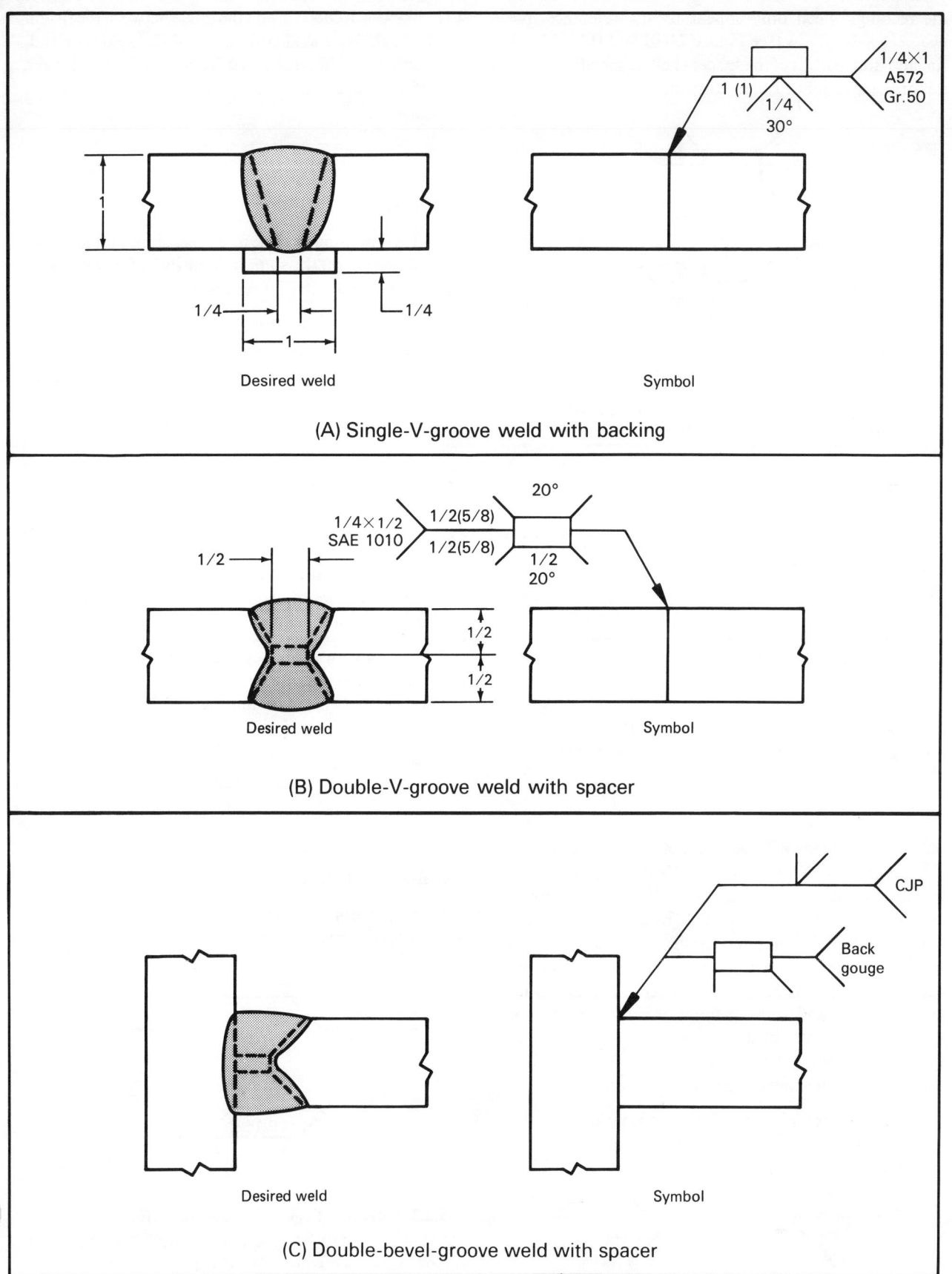

Figure 24 — Joints with Backing and Spacers

the rectangle need only appear on the reference line nearest to the arrow [see Figure 24(B) and (C)]. Material and dimensions of the spacer shall be specified in the tail of the symbol or on the drawing.

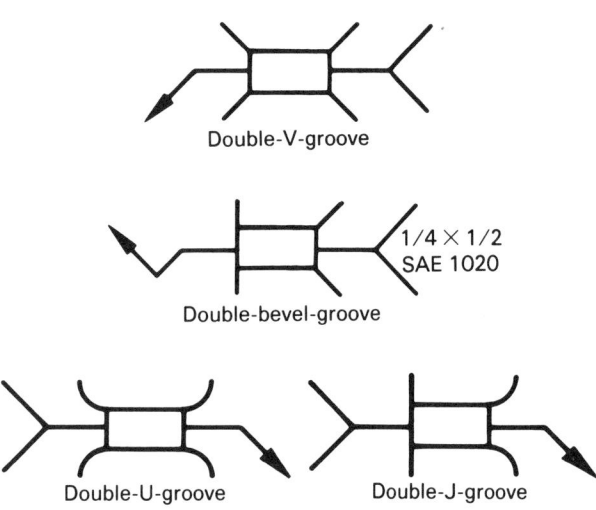

4.8 Consumable Inserts. Consumable inserts shall be specified by placing the consumable insert symbol on the side of the reference line opposite the groove weld symbol. The AWS consumable insert class shall be placed in the tail of the symbol (for insert class see latest edition of AWS A5.30).

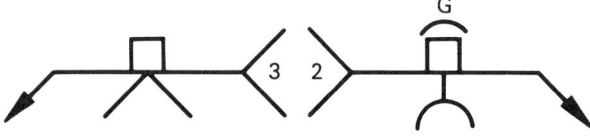

4.9 Groove Welds with Back Gouging. A joint requiring complete penetration involving back gouging may be specified using either a single or multiple reference line symbol. That welding symbol shall include a reference to back gouging in its tail and (1) in the case of assymetrical groove welds must show the depth of preparation from each side [see Figure 25(A) and (B)], together with groove angles and root opening, or (2) in the case of symetrical groove welds, need not include any other information except the weld symbol [see 4.2.2 and Figure 25(C)], with groove angles and root opening.

4.10 Seal Welds. When the intent of the weld is to fulfill a sealing function only, the weld shall be specified in the tail of the symbol as a seal weld.

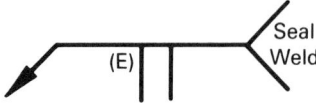

4.11 Skewed Joints. When the angle between the fusion faces is such that the identification of the weld type and, hence, proper weld symbol is in question, the detail of the desired joint and weld configuration shall be shown on the drawing.

5. Fillet Welds

5.1 General

5.1.1 Dimension Location. Dimensions of fillet welds shall be shown on the same side of the reference line as the weld symbol (see Figures 26–28).

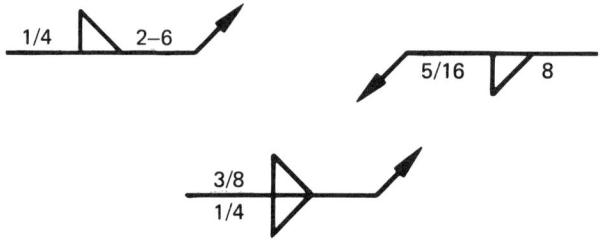

5.1.2 Double Fillet Welds. The dimensions of fillet welds on both sides of a joint shall be specified whether the dimensions are identical or different.

5.1.3 Drawing Notes. Dimensions of fillet welds covered by drawing notes need not be repeated on the welding symbols in accordance with 3.10.6.

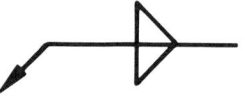

5.2 Size of Fillet Welds

5.2.1 Location. The fillet weld size shall be specified to the left of the weld symbol (see Figure 26).

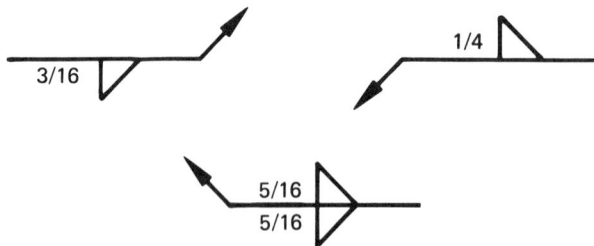

5.2.2 Unequal Legs. The size of a fillet weld with unequal legs shall be specified to the left of the weld symbol as shown below. Weld orientation is not speci-

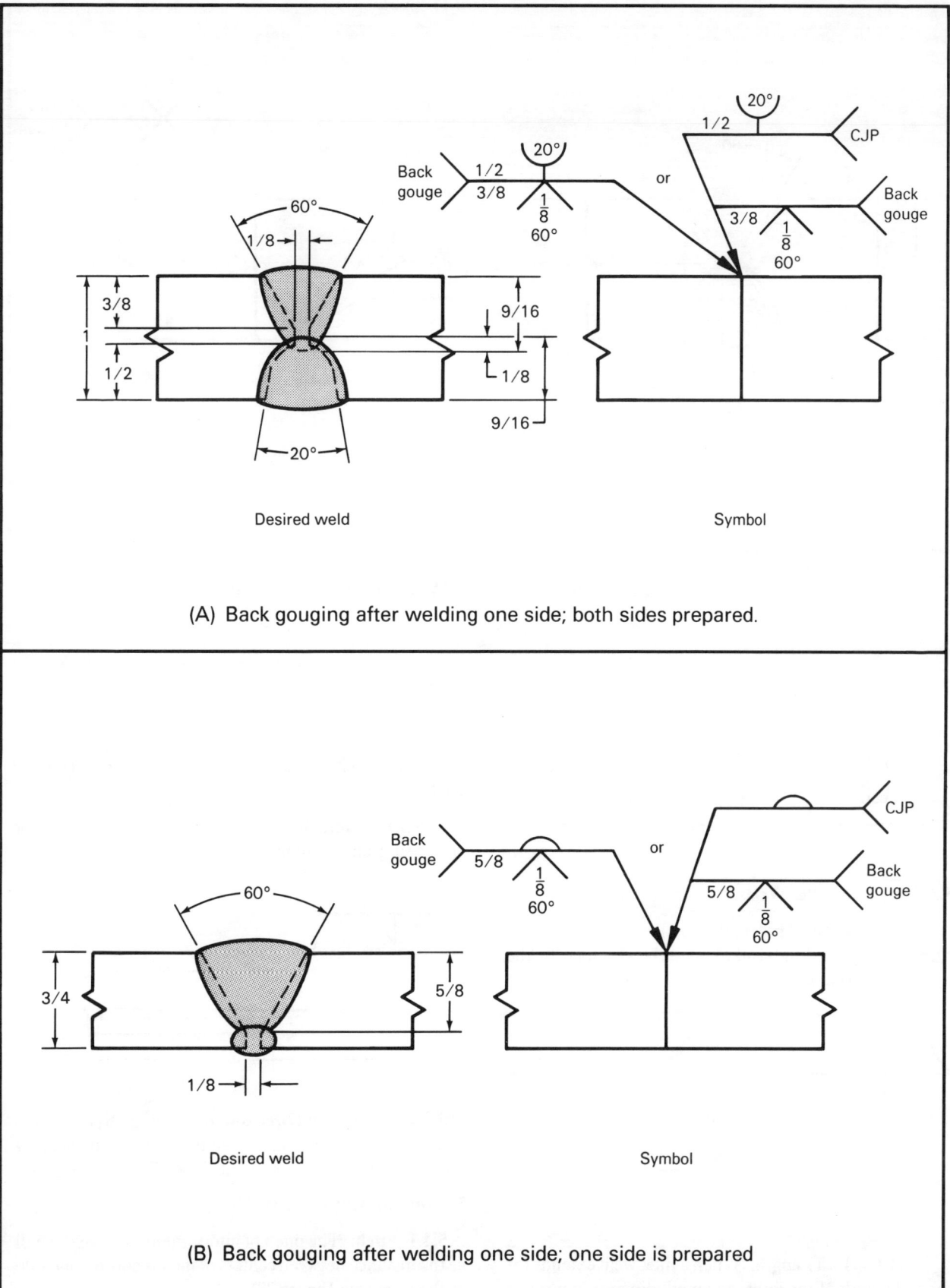

Figure 25 — Groove Welds with Back Gouging

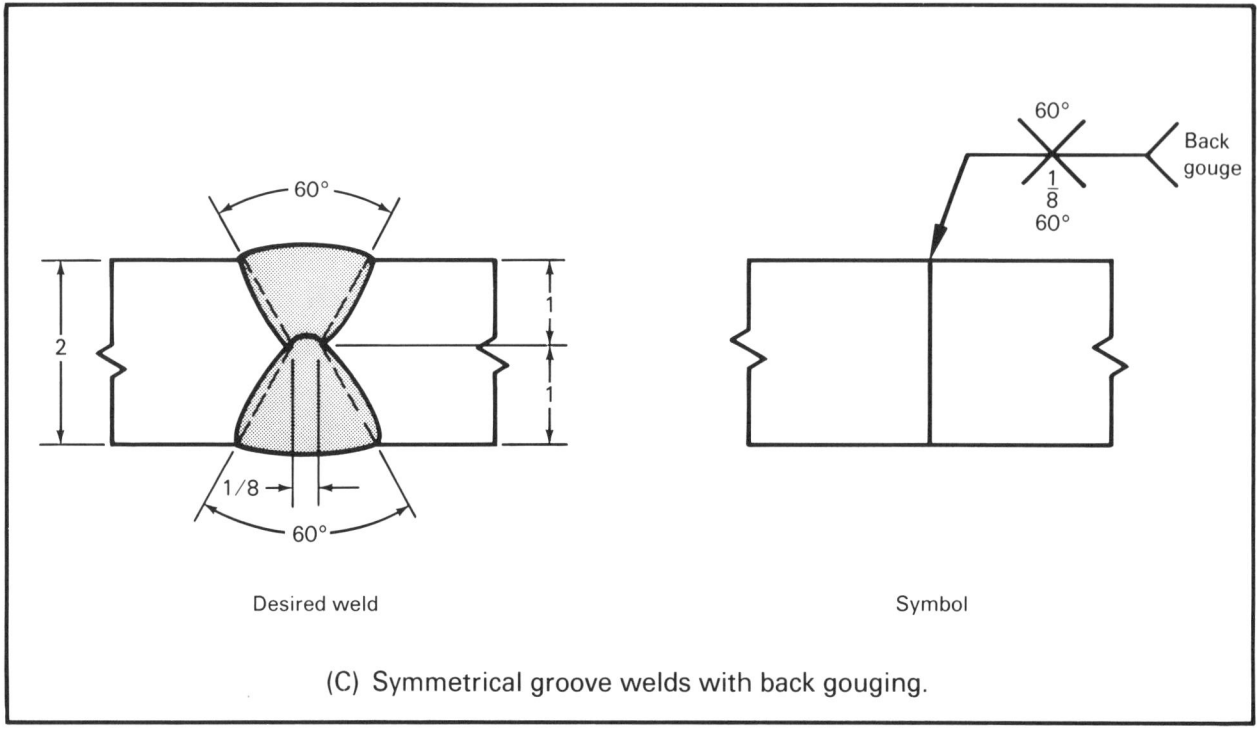

(C) Symmetrical groove welds with back gouging.

Figure 25 (continued) — Groove Welds with Back Gouging

fied by the symbol and shall be shown on the drawing to ensure clarity [see Figure 26(D)].

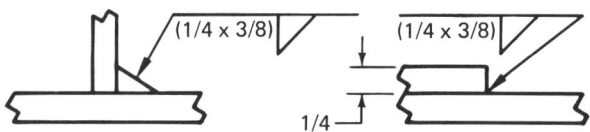

5.3 Length of Fillet Welds

5.3.1 Location. The length of a fillet weld, when indicated on the welding symbol, shall be specified to the right of the weld symbol [see Figure 26(F)].

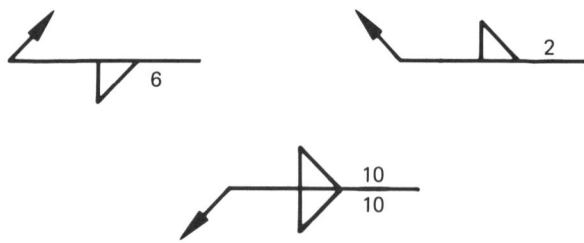

5.3.1.1 Full Length. When a fillet weld extends for the full length of the joint, no length dimension need be specified on the welding symbol [see Figure 26(A), (B), (C), (D) and (E)].

5.3.1.2 Specific Lengths. Specific lengths of fillet welds, and their location, may be specified by symbols in conjunction with dimension lines [see Figures 8(C) and 26(F)].

5.3.1.3 Hatching. The extent of fillet welds may be specified graphically by hatching.

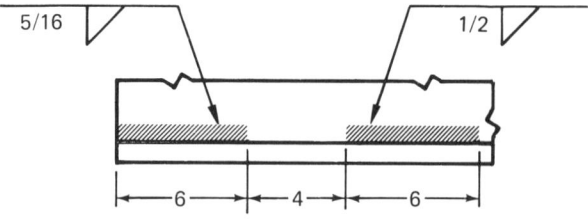

5.3.2 Changes in Direction of Welding. Symbols for fillet welds involving changes in the direction of welding shall be in accordance with 3.8.2 [see Figure 9(A)].

5.4 Intermittent Fillet Welds

5.4.1 Pitch. The pitch of intermittent fillet welds shall be the distance between centers of increments on one side of the joint (see Figure 27).

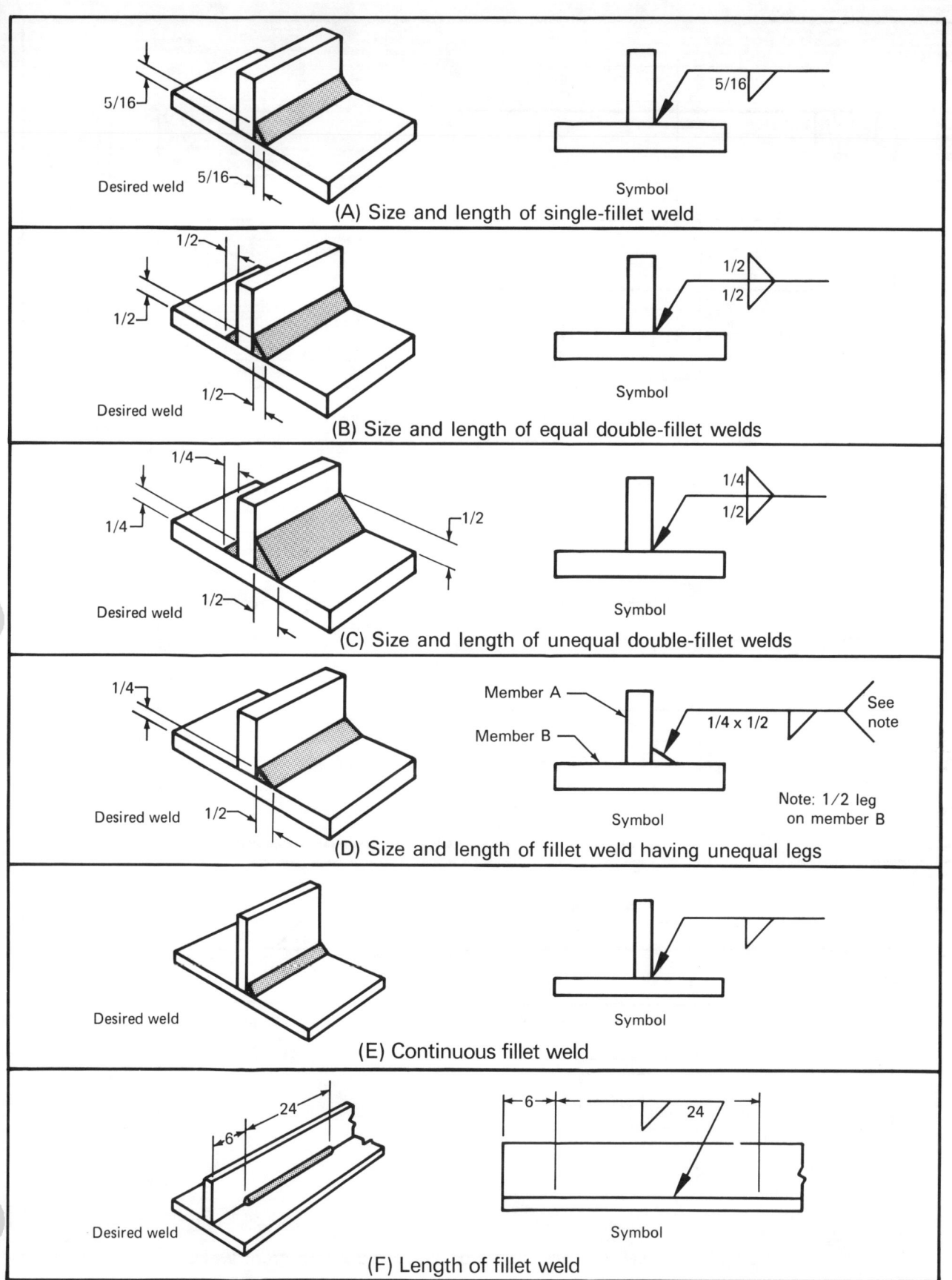

Figure 26 — Application of Dimensions to Fillet Weld Symbols

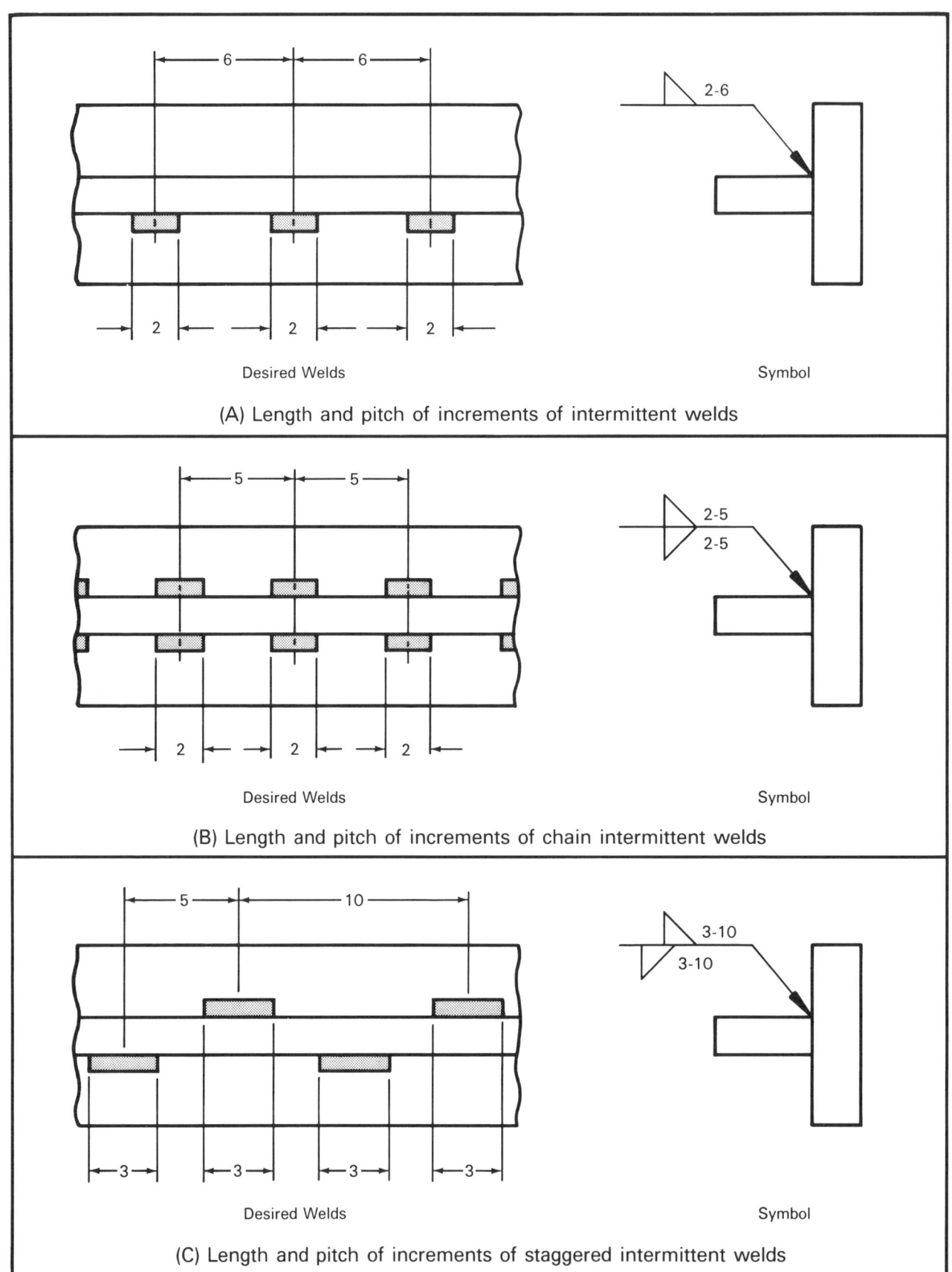

Figure 27 — Application of Dimensions to Intermittent Fillet Weld Symbols

5.4.2 Pitch Dimension Location. The pitch of intermittent fillet welds shall be specified to the right of the length dimension following a hyphen (see Figure 27).

5.4.3 Chain Intermittent Fillet Welds. Dimensions of chain intermittent fillet welds shall be specified on both sides of the reference line. The segments of chain intermittent fillet welds shall be opposite one another across the joint [see Figure 27(B)].

5.4.4 Staggered Intermittent Fillet Welds. Dimensions of staggered intermittent fillet welds shall be specified on both sides of the reference line, and the fillet weld symbols shall not be directly opposite across the reference line. The segments of staggered intermittent fillet welds shall be symmetrically spaced on both sides of the joint as shown in Figure 27(C).

5.4.5 Extent of Welding. In the case of intermittent fillets, when additional welds are intended at the ends of the joint, they should be specified by separate symbols and dimensioned on the drawings. When no welds are intended at the ends of the joint, the unwelded length should not exceed that specified by the symbol and should be so dimensioned on the drawing.

5.5 Fillet Welds in Holes and Slots. Fillet welds in holes and slots shall be specified by the use of fillet weld symbols [see Figure 28(A)].

5.6 Contours and Finishing of Fillet Welds

5.6.1 Contours Obtained by Welding. Fillet welds that are to be welded with approximately flat, convex, or concave faces without postweld finishing shall be specified by adding the flat, convex, or concave contour symbol to the weld symbol in accordance with the location conventions given in 3.1.

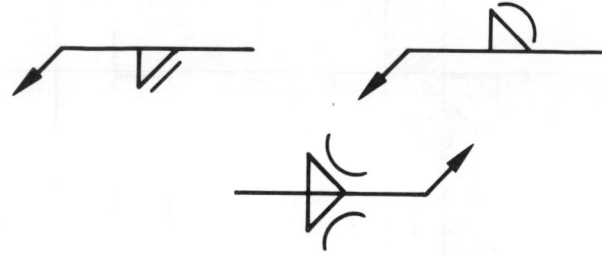

5.6.2 Contours Obtained by Postweld Finishing. Fillet welds whose faces are to be finished approximately flat, convex, or concave by postweld finishing shall be specified by adding both the appropriate contour and finishing symbol to the weld symbol (see 3.12.1).

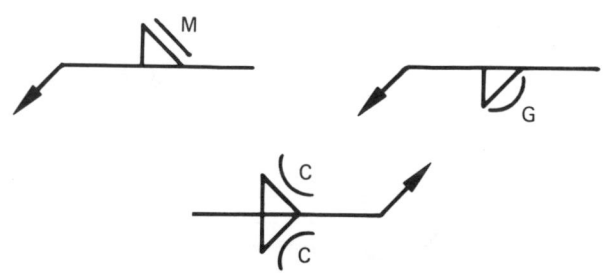

5.7 Skewed Joints. When the angle between the fusion faces is such that the identification of the weld type and, hence, proper weld symbol is in question, the detail of the desired joint and weld configuration shall be shown on the drawing.

6. Plug Welds

6.1 General

6.1.1 Arrow-Side Holes. Holes in the arrow-side member of a joint to be plug welded shall be specified by placing the weld symbol below the reference line [see Figure 29(A)].

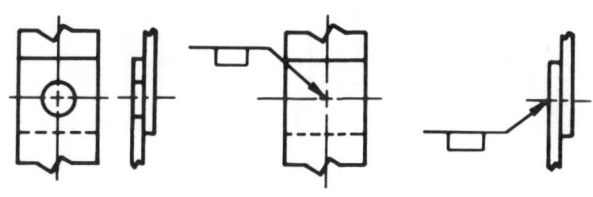

Desired   Symbols

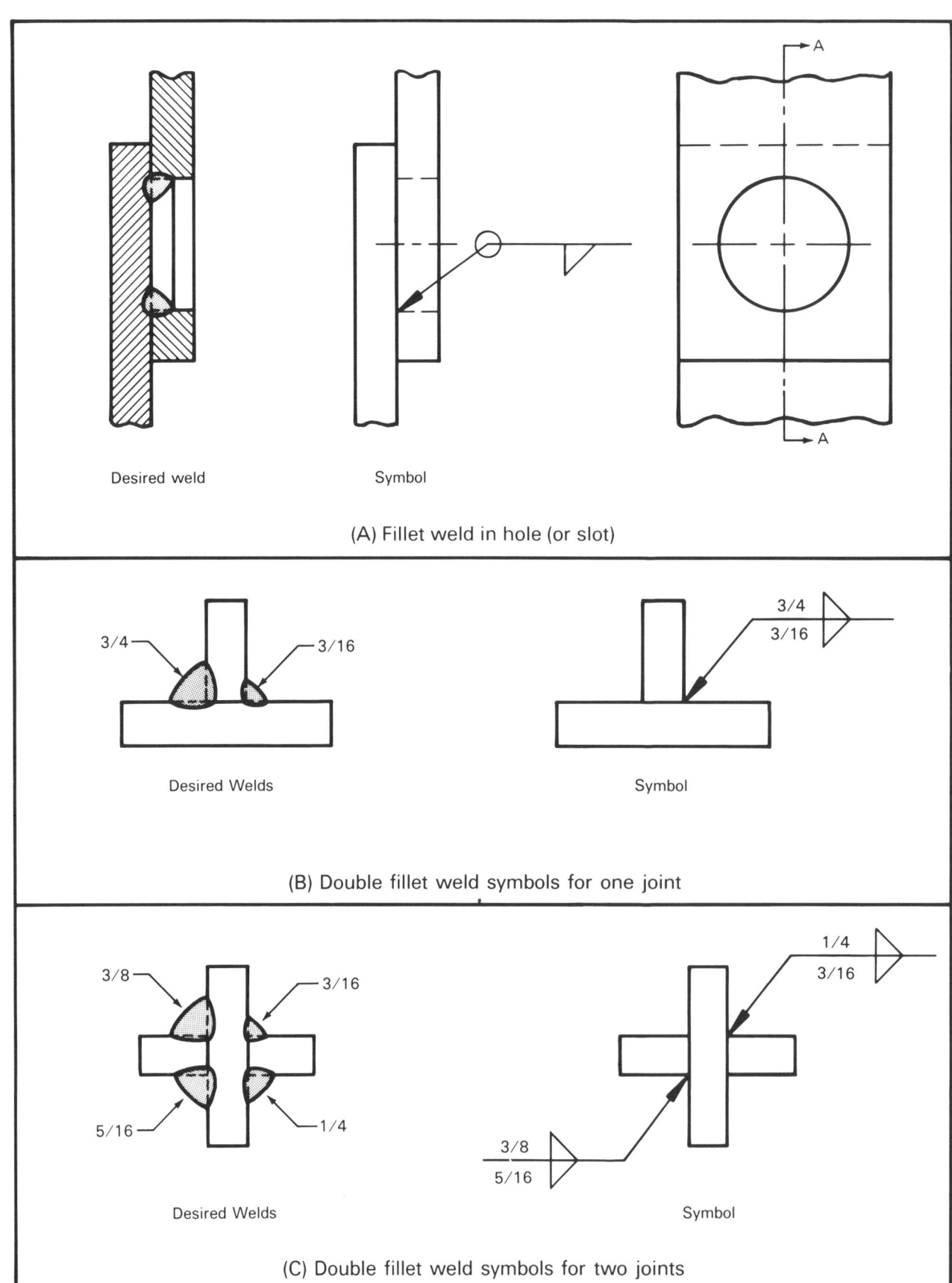

Figure 28 — Application of Fillet Weld Symbols

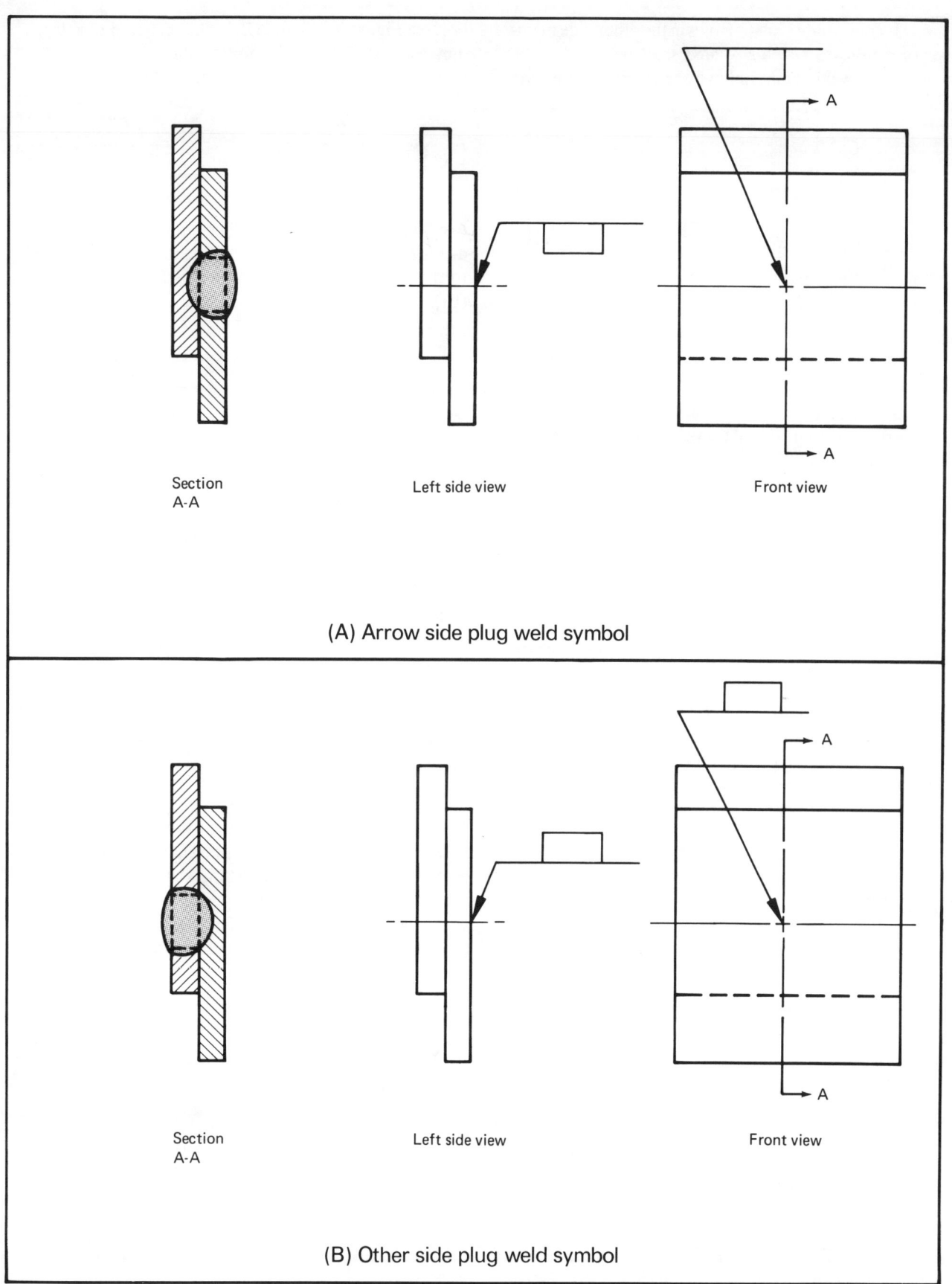

Figure 29 — Application of Plug Weld Symbol

6.1.2 Other-Side Holes. Holes in the other-side member of a joint to be plug welded shall be specified by placing the weld symbol above the reference line [see Figure 29(B)].

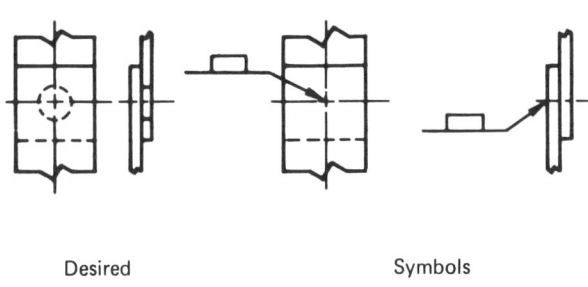

Desired   Symbols

6.1.3 Dimensions. Dimensions of plug welds shall be specified on the same side of the reference line as the weld symbol (see Figure 30).

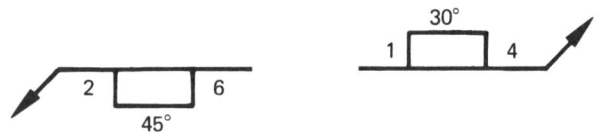

6.1.4 Fillets in Holes. The plug weld symbol shall not be used to designate fillet welds in holes (see 5.5).

6.2 Size of Plug Welds. The size of a plug weld shall be specified to the left of the weld symbol [see Figure 30(A), (B), (E), (F), and (G)]. Plug weld size is the diameter of the hole at the faying surface.

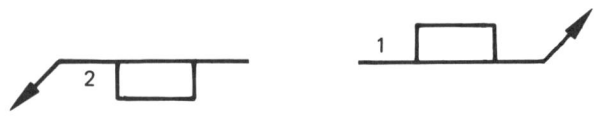

6.3 Angle of Countersink. The included angle of countersink of plug welds shall be specified either below or above the plug weld symbol [see Figure 30(B), (E), (F), and (G)].

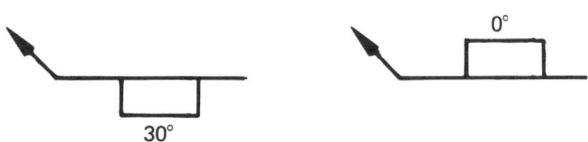

6.4 Depth of Filling. When the depth of filling is less than complete, it shall be specified inside the weld symbol [see Figure 30(C) and (E)]. The omission of a depth dimension shall specify complete filling.

6.5 Spacing of Plug Welds. The pitch (center-to-center distance) of plug welds in a straight line shall be specified to the right of the weld symbol [see Figure 30(D) and (E)].

The spacing of plug welds in any configuration other than a straight line shall be dimensioned on the drawing.

6.6 Contours and Finishing of Plug Welds

6.6.1 Contours Obtained by Welding. Plug welds that are to be welded with approximately flush or convex faces without postweld finishing shall be specified by adding the flush or convex contour symbol to the weld symbol.

6.6.2 Contours Obtained by Postweld Finishing. Plug welds whose faces are to be finished approximately flush or convex by post weld finishing shall be specified by adding both the appropriate contour and finishing symbol to the welding symbol (see 3.12.1). Welds that require a flat but not flush surface, require an explanatory note in the tail of the symbol.

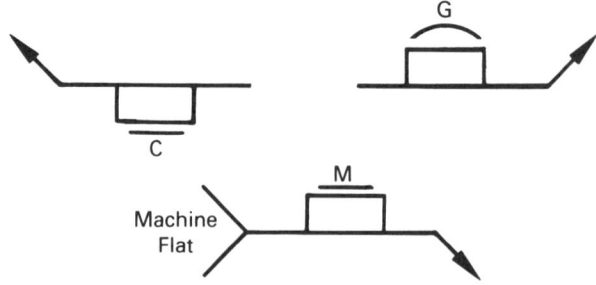

6.7 Joints Involving Three or More Members. Plug welding symbols may be used to specify welding two or

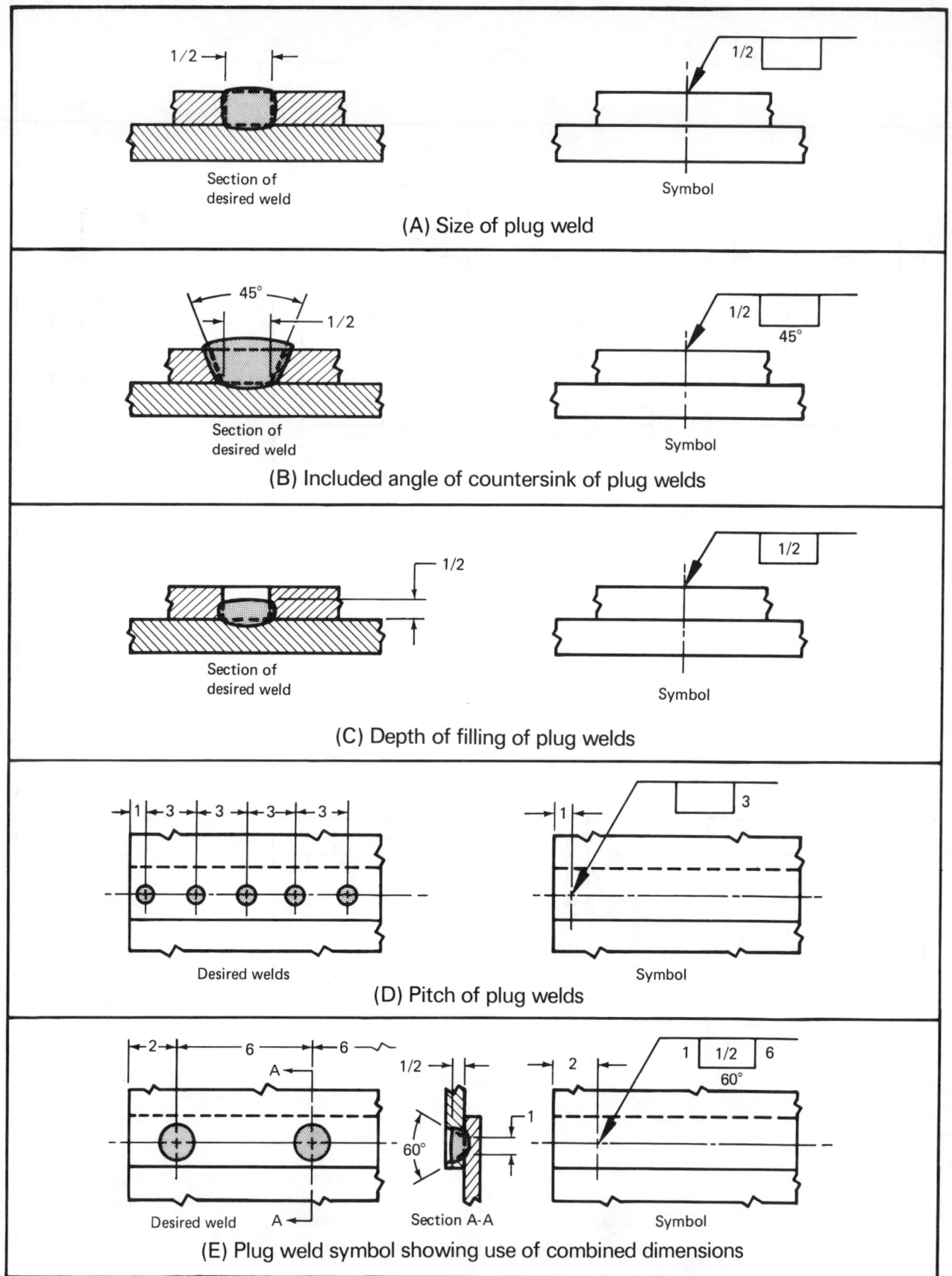

Figure 30 — Application of Dimensions to Plug Weld Symbols

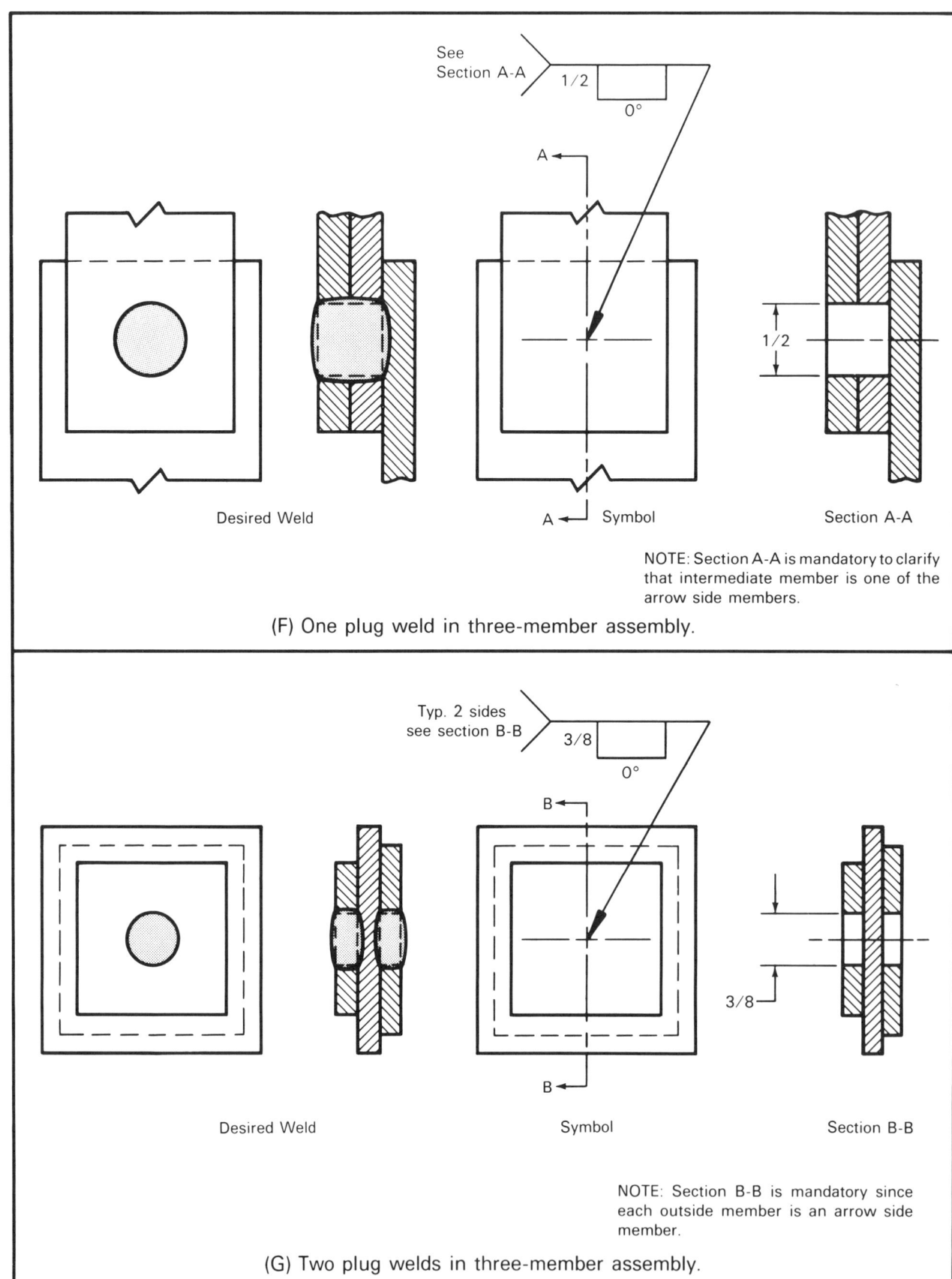

Figure 30 (continued) — Application of Dimensions to Plug Weld Symbols

more members to another member. A section view of the joint shall be provided to clarify which members require preparation [see Figure 30(F) and (G)].

7. Slot Welds

7.1 General

7.1.1 Arrow-Side Slots. Slots in the arrow-side member of a joint to be slot welded shall be specified by placing the weld symbol below the reference line [see Figure 31(A)].

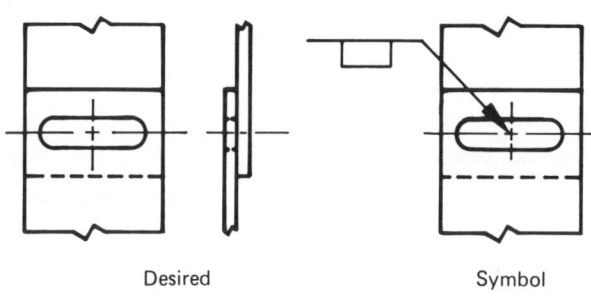

Desired   Symbol

7.1.2 Other-Side Slots. Slots in the other-side member of a joint to be slot welded shall be specified by placing the weld symbol above the reference line [see Figure 31(B)].

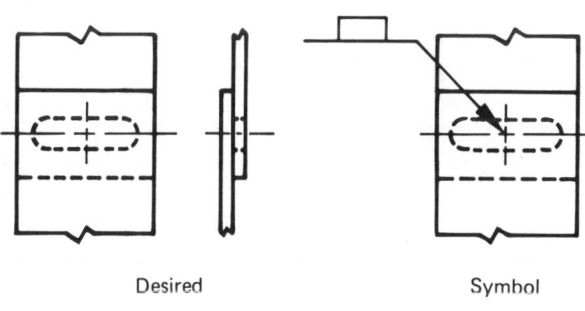

Desired   Symbol

7.1.3 Fillets in Slots. The slot weld symbol shall not be used to specify fillet welds in slots (see 5.5).

7.2 Depth of Filling. Depth of filling less than complete shall be specified inside the weld symbol (see Figure 32). Omission of depth dimension shall specify complete filling.

7.3 Details of Slot Welds. Length, width, spacing, angle of bevel, orientation, and location of slot welds cannot be specified on the welding symbol. These data shall be specified on the drawing, or by a detail with a reference thereto on the welding symbol (see Figure 32).

7.4 Contours and Finishing of Slot Welds

7.4.1 Contours Obtained by Welding. Slot welds that are to be welded with approximately flush or convex faces without postweld finishing shall be specified by adding the flush or convex contour symbol to the weld symbol.

7.4.2 Contours Obtained by Postweld Finishing. Spot welds whose faces are to be finished approximately flush or convex by postweld finishing shall be specified by adding both the appropriate contour and finishing symbol to the welding symbol (see 3.12.1). Welds that require a flat but not flush surface require an explanatory note in the tail of the symbol.

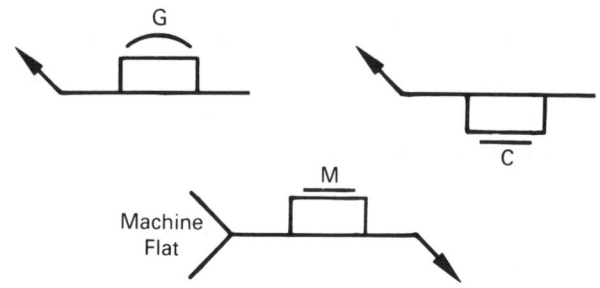

8. Spot Welds

8.1 General

8.1.1 Arrow-Side Significance. The spot weld symbol, relative to its location on the reference line, may or may not have arrow-side member or other-side member significance.

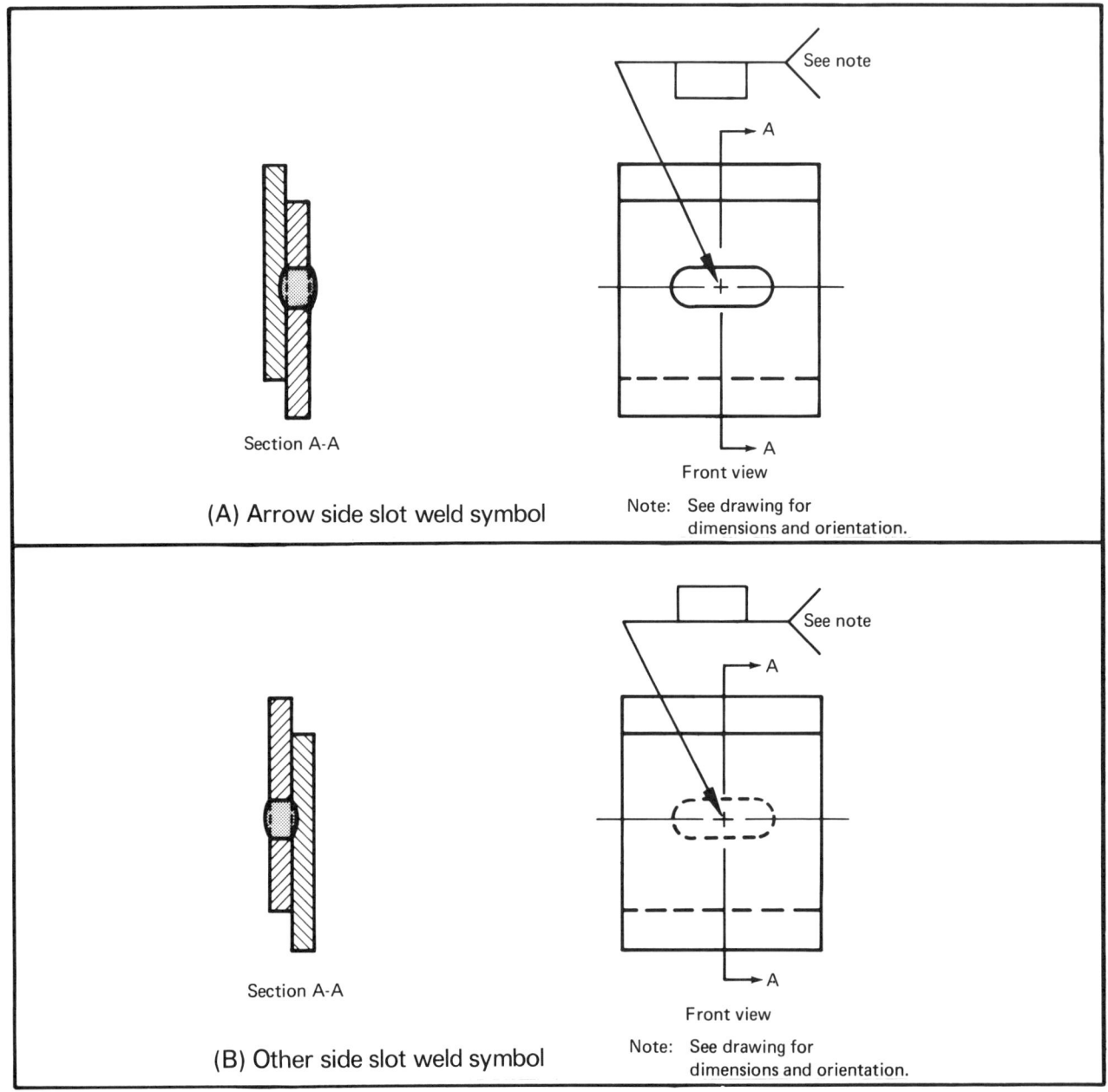

Figure 31 — Application of Slot Weld Symbol

8.1.1.1 Arrow-Side Member. For those welding processes for which arrow-side member significance is applicable, the arrow-side member shall be indicated by placing the spot weld symbol below the reference line with the arrow pointing to this member [see Figure 33 (A)].

8.1.1.2 Other-Side Member. For those welding processes for which other-side member significance is applicable, the other-side member shall be indicated by placing the weld symbol above the reference line [see Figure 33(B)].

8.1.2 Dimension Location. Dimensions shall be specified on the same side of the reference line as the symbol, or all dimensions on either side when the symbol is

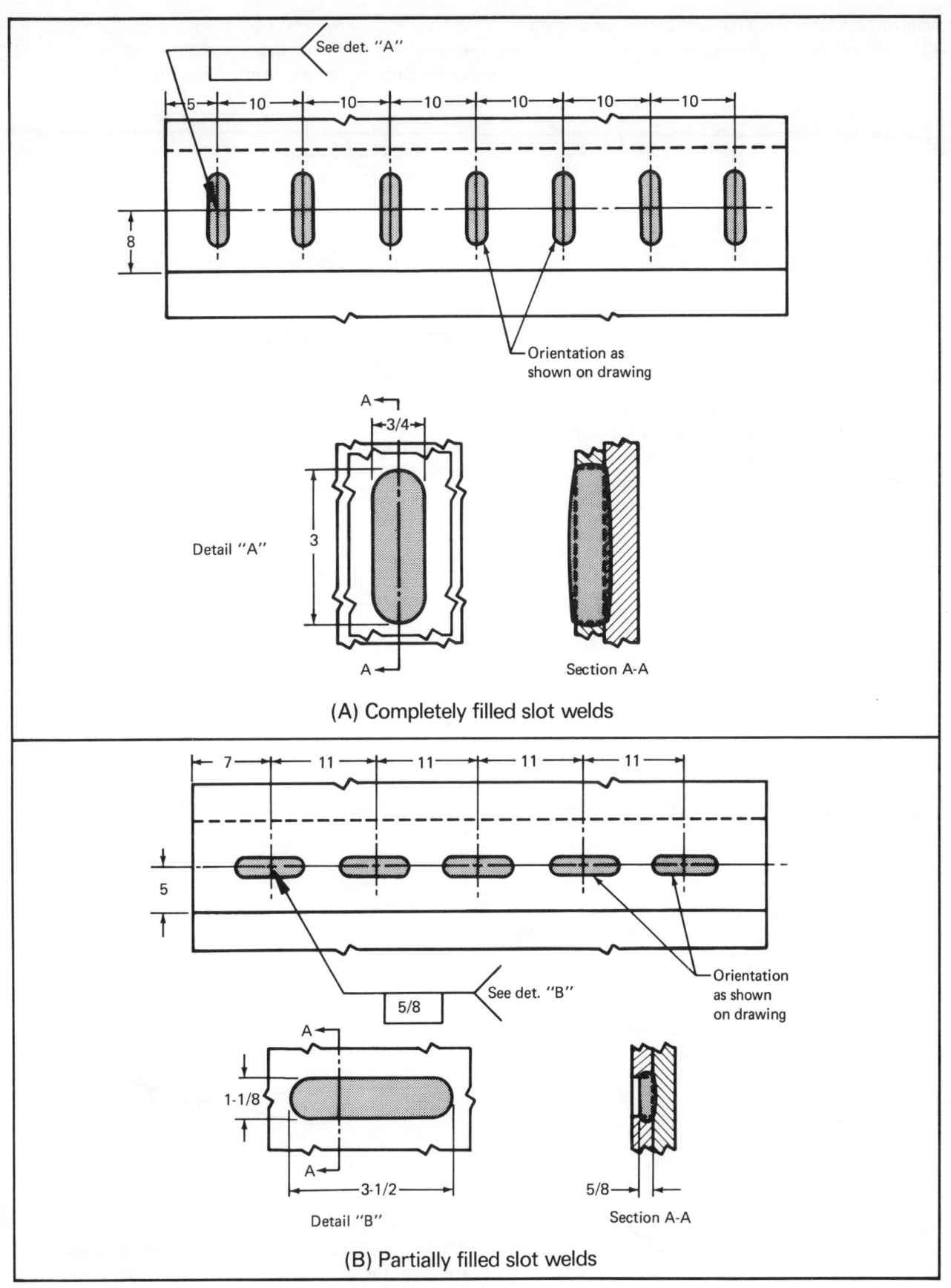

Figure 32 — Application of Dimensions to Slot Weld Symbols

centered on the reference line and has no arrow-side or other-side significance (see Figure 33).

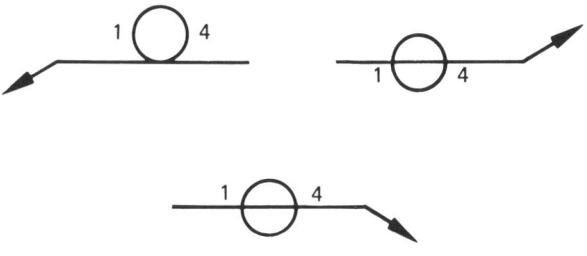

8.1.3 Process Reference. The process reference shall be indicated in the tail of the welding symbol (see 3.10.2 and Figure 33).

8.1.4 Projection Welds. When projection welding is to be employed, the projection weld symbol shall be used with the projection welding process reference in the tail of the welding symbol. The projection weld symbol shall be centered above or below (not on) the reference line to designate in which member the embossment is placed in accordance with the location conventions given in 3.1.2 (see Figure 34).

8.2 Size and Strength of Spot Welds. Spot welds shall be dimensioned by either size or strength as follows:

8.2.1 Size. The size of spot welds may be specified as the diameter of the weld at the faying surfaces of the members [see Figure 35(A)].

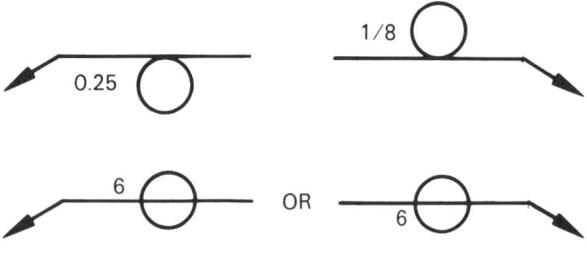

8.2.2 Strength. The shear strength of spot welds may be specified in pounds or newtons per spot, to the left of the weld symbol (see Figure 35(B)].

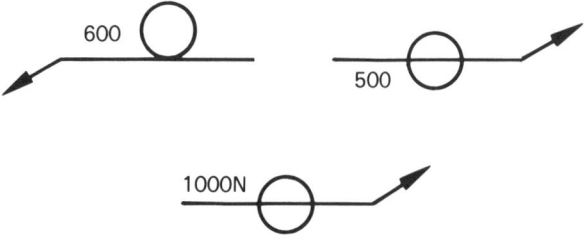

8.3 Spacing of Spot Welds. The pitch (center-to-center distance) of spot welds in a straight line shall be specified to the right of the weld symbol [see Figure 35(C)].

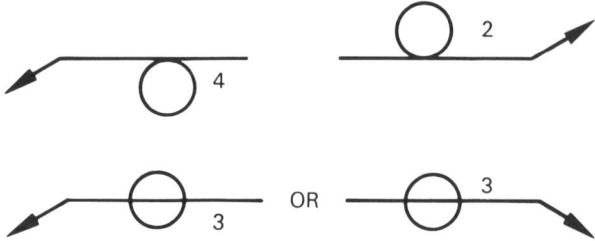

8.4 Number of Spot Welds

8.4.1 Number Specified. When a definite number of spot welds is desired in a joint, the number shall be specified in parentheses on the same side of the reference line as that of the weld symbol. The number may be either above or below the weld symbol, as appropriate (see Figure 35(C), (D), (E), and (F)].

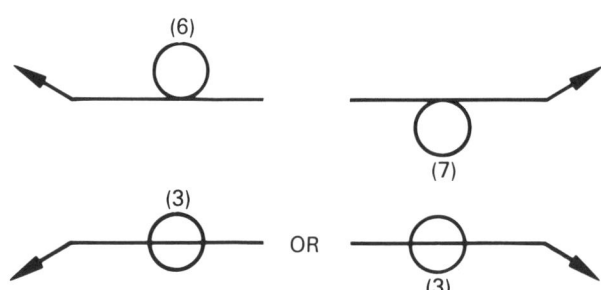

8.4.2 Grouped Spot Welds. A group of spot welds may be located on a drawing by intersecting center lines. The arrow shall point to at least one of the center lines passing through each weld location.

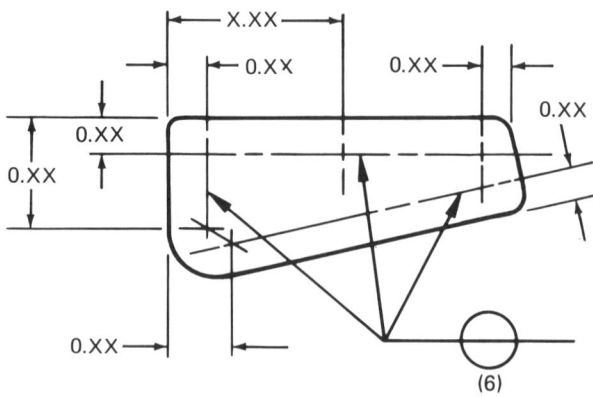

8.5 Extent of Spot Welding. When spot welding extends less than the distance between abrupt changes in the

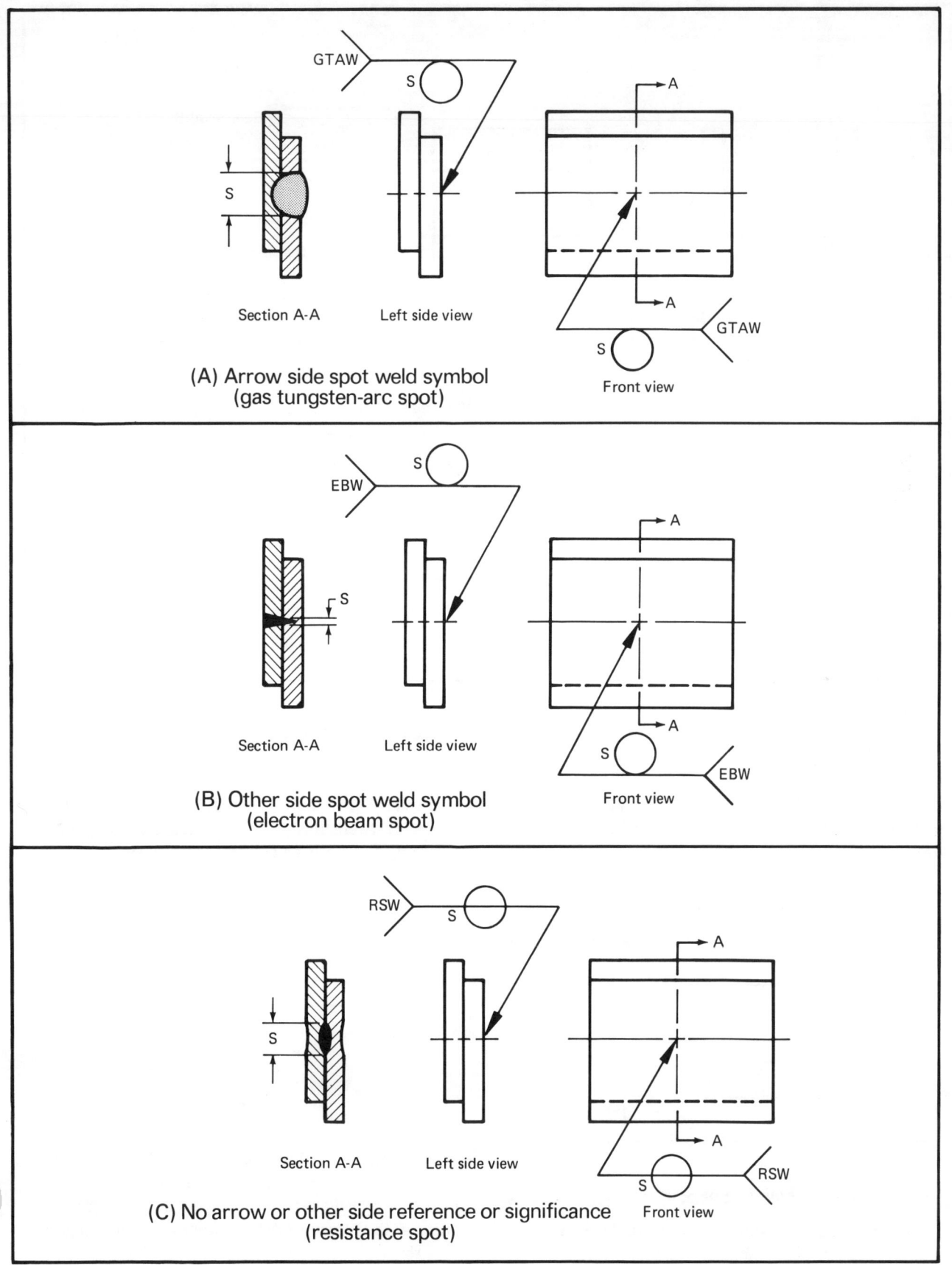

Figure 33 — Application of Spot Weld Symbol

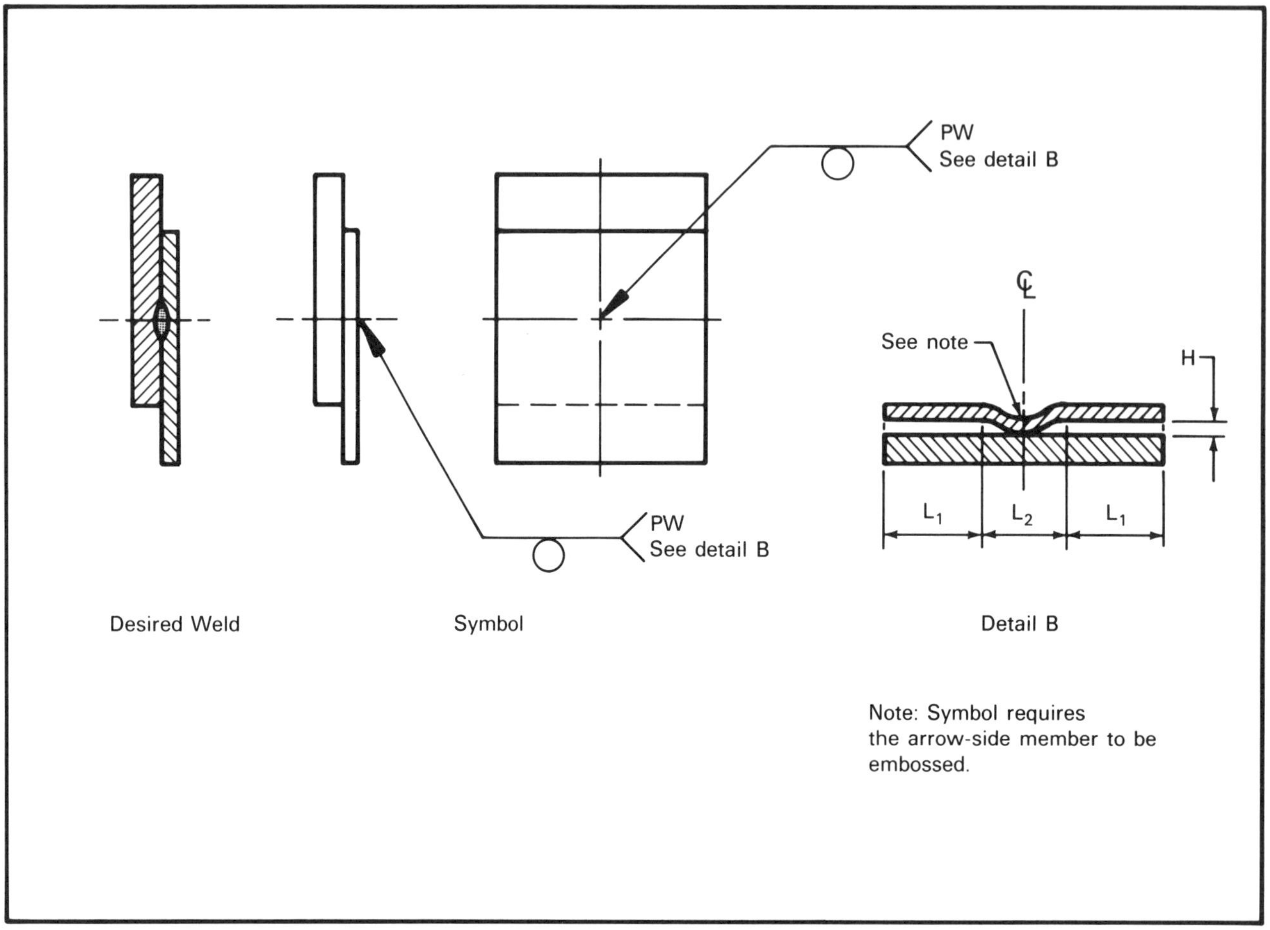

Figure 34 — Application of Projection Weld Symbol

direction of the welding, or less than the full length of the joint (see 3.8), the extent shall be dimensioned (see Figure 35(D)].

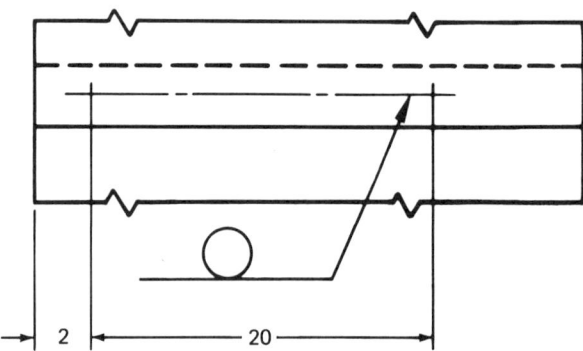

8.6 Contours and Finishing of Spot Welds

8.6.1 Contours Obtained by Welding. When the exposed surface of either member of a spot welded joint is to be welded with approximately flush or convex faces without postweld finishing, that surface shall be specified by adding the flush or convex contour symbol to the weld symbol.

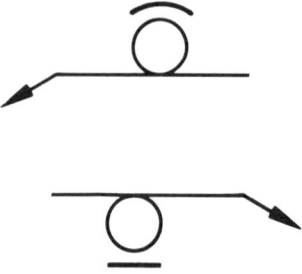

8.6.2 Contours Obtained by Postweld Finishing. Spot welds whose faces are to be finished approximately flush, or convex by postweld finishing, shall be specified by adding both the appropriate contour and finishing symbol to the welding symbol (see 3.12.1). Welds that require

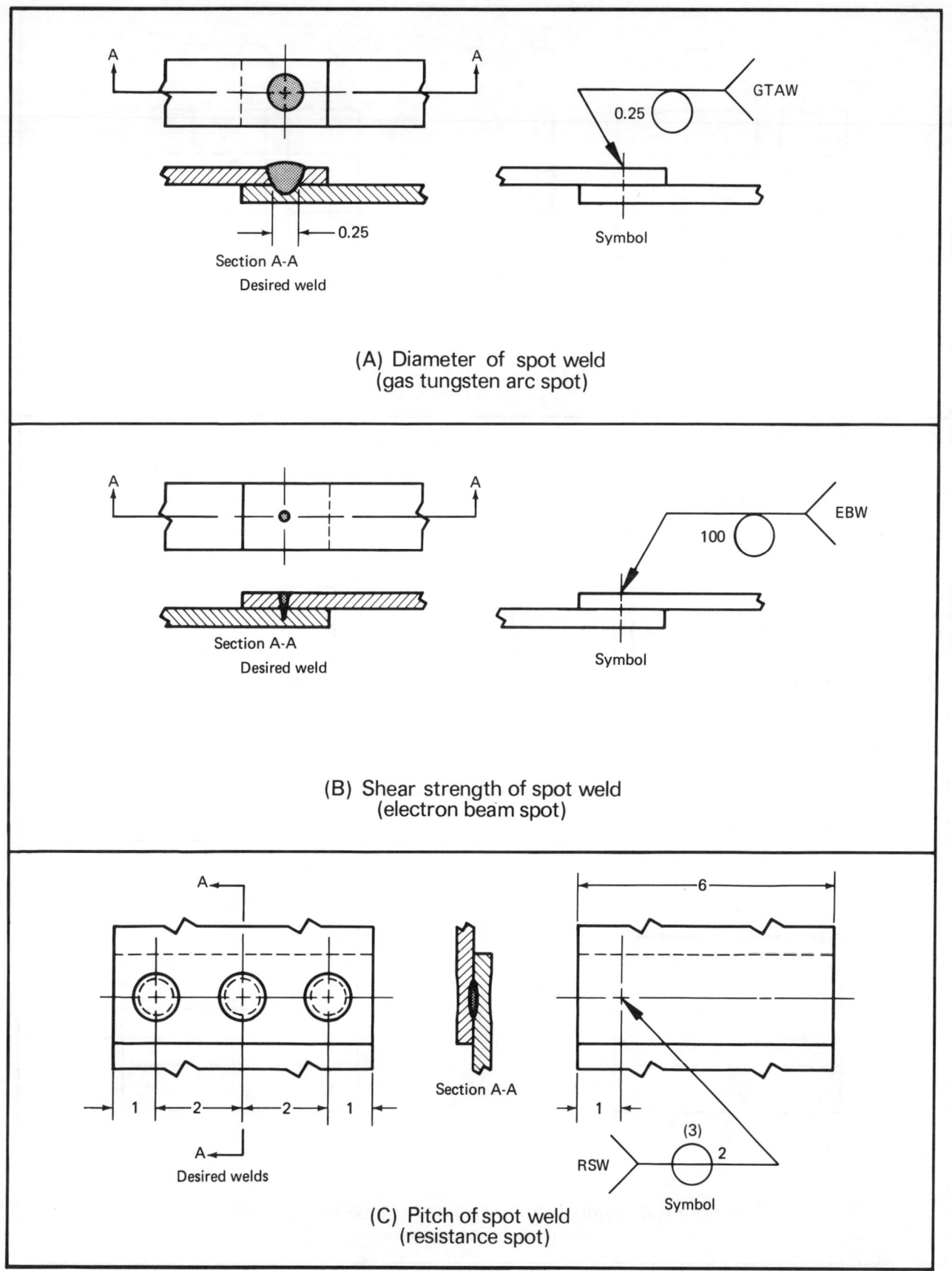

Figure 35 — Application of Dimensions to Spot Weld Symbols

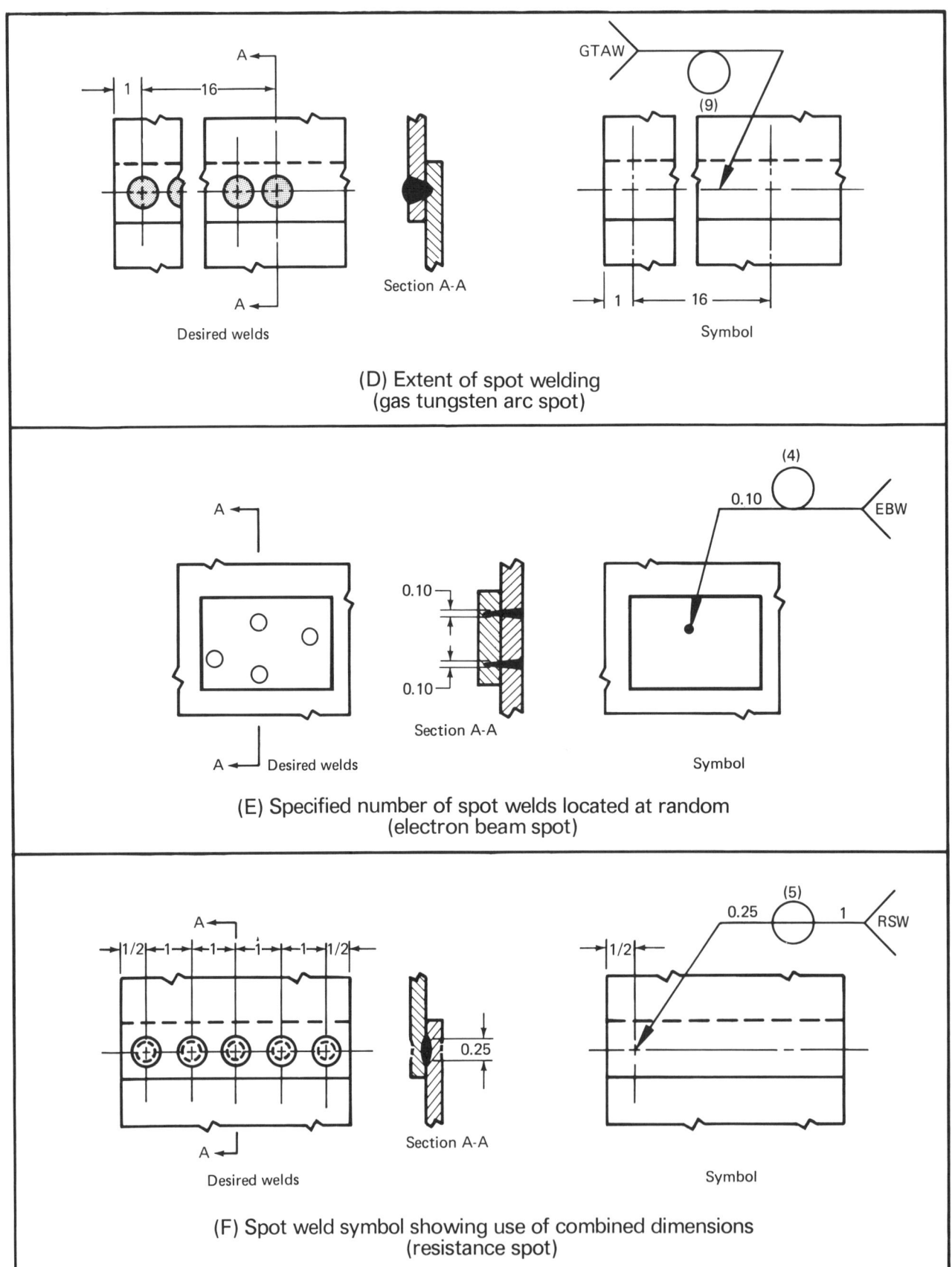

Figure 35 (continued) — Application of Dimensions to Spot Weld Symbols

a flat but not flush surface require an explanatory note in the tail of the symbol.

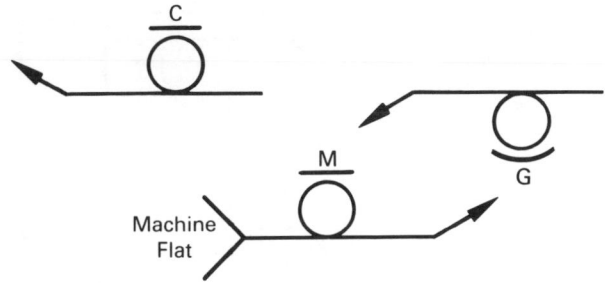

8.7 Multiple-Joint Spot Welds. When one or more members are included between the two outer members in a spot welded joint, the spot weld symbol shall be used.

9. Seam Welds

9.1 General

9.1.1 Arrow-Side Significance. The seam weld symbol, relative to its location on the reference line, may or may not have arrow-side or other significance.

9.1.1.1 Arrow-Side Member. For those welding processes for which arrow-side member significance is applicable, the arrow-side member shall be indicated by placing the seam weld symbol below the reference line with the arrow pointing to this member [see Figure 36(A)].

9.1.1.2 Other-Side Member. For those welding processes for which other-side significance is applicable, the other-side member shall be indicated by placing the weld symbol above the reference line [see Figure 36(B)].

9.1.2 Dimensions. Dimensions shall be shown on the same side of the reference line as the symbol, or all on either side when the symbol is located centered on the reference line and has no arrow-side or other-side significance (see Figure 36).

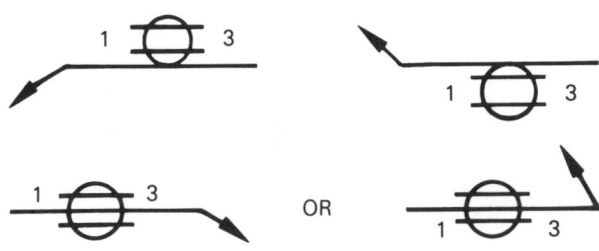

9.1.3 Process Reference. The process reference shall be indicated in the tail of the welding symbol. (See 3.10.1 and Figure 36).

9.2 Size and Strength of Seam Welds. Seam welds shall be dimensioned by either size or strength as follows:

9.2.1 Size. When size of seam welds is required, it shall be specified as the width of the weld at the faying surfaces of the members and located to the left of the weld symbol [see Figure 37(A)].

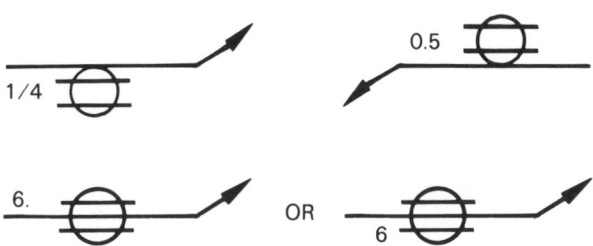

9.2.2 Strength. When instead of size, the shear strength of seam welds is required it shall be specified in pounds per linear inch or in newtons per millimeter, and located to the left of the weld symbol [see Figure 37(B)].

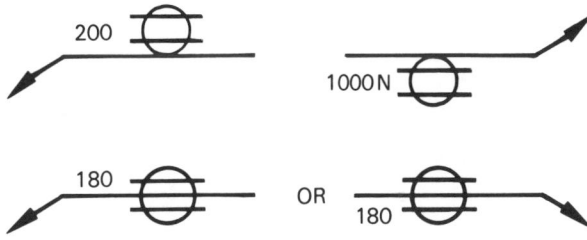

9.3 Length of Seam Welds

9.3.1 Dimension Location. The length of seam welds, when indicated on welding symbols, shall be specified to the right of the weld symbol [see Figure 37(A)].

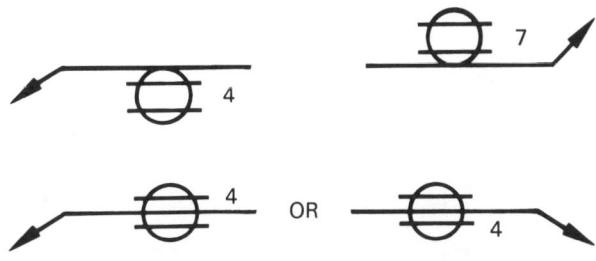

9.3.2 Abrupt Changes. When a seam weld extends the full distance between abrupt changes in the direction of the welding (see 3.8), no length dimension need be specified on the welding symbol.

9.3.3 Specific Lengths. When a seam weld extends less than the distance between abrupt changes in the

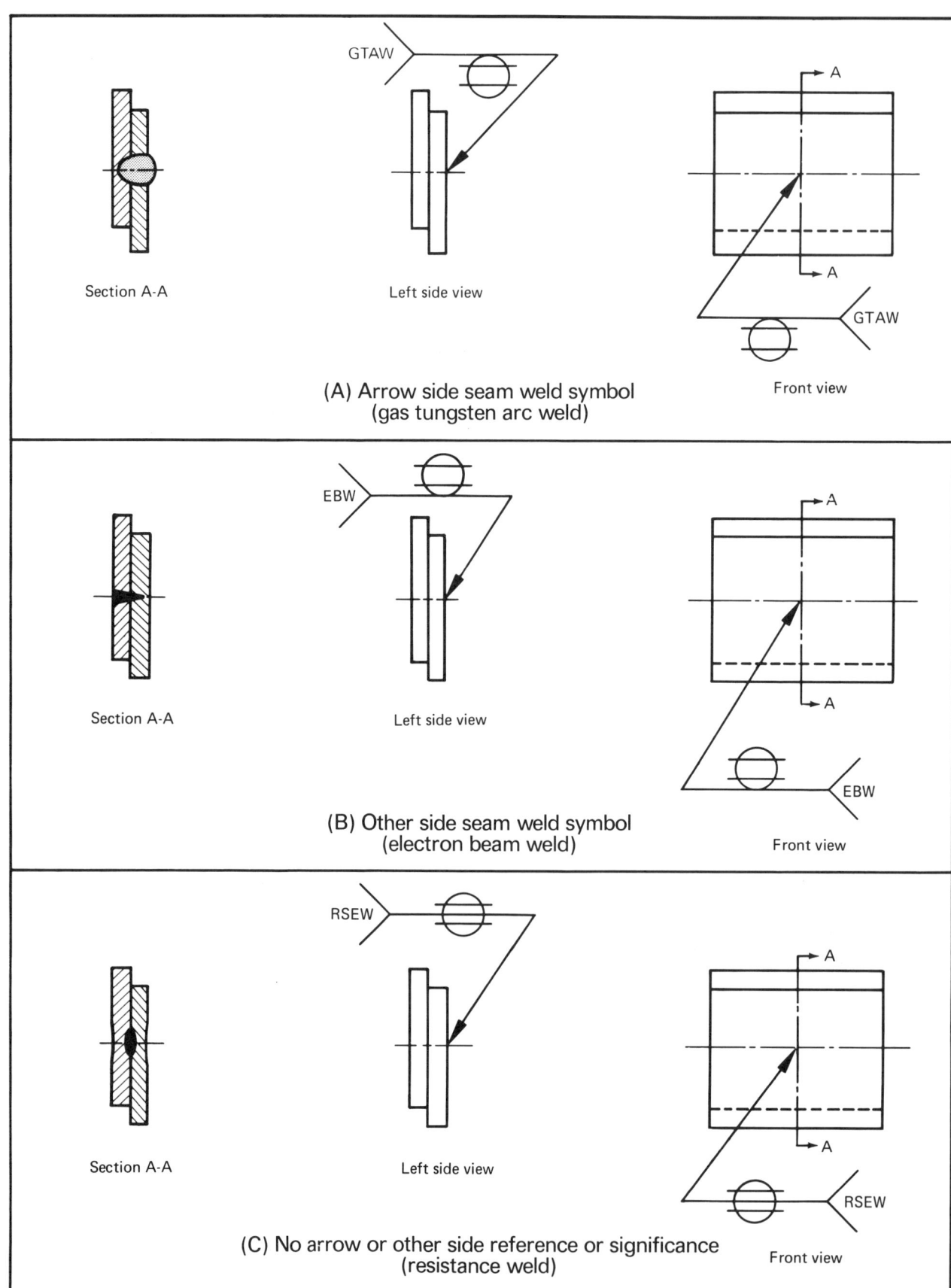

Figure 36 — Application of Seam Weld Symbol

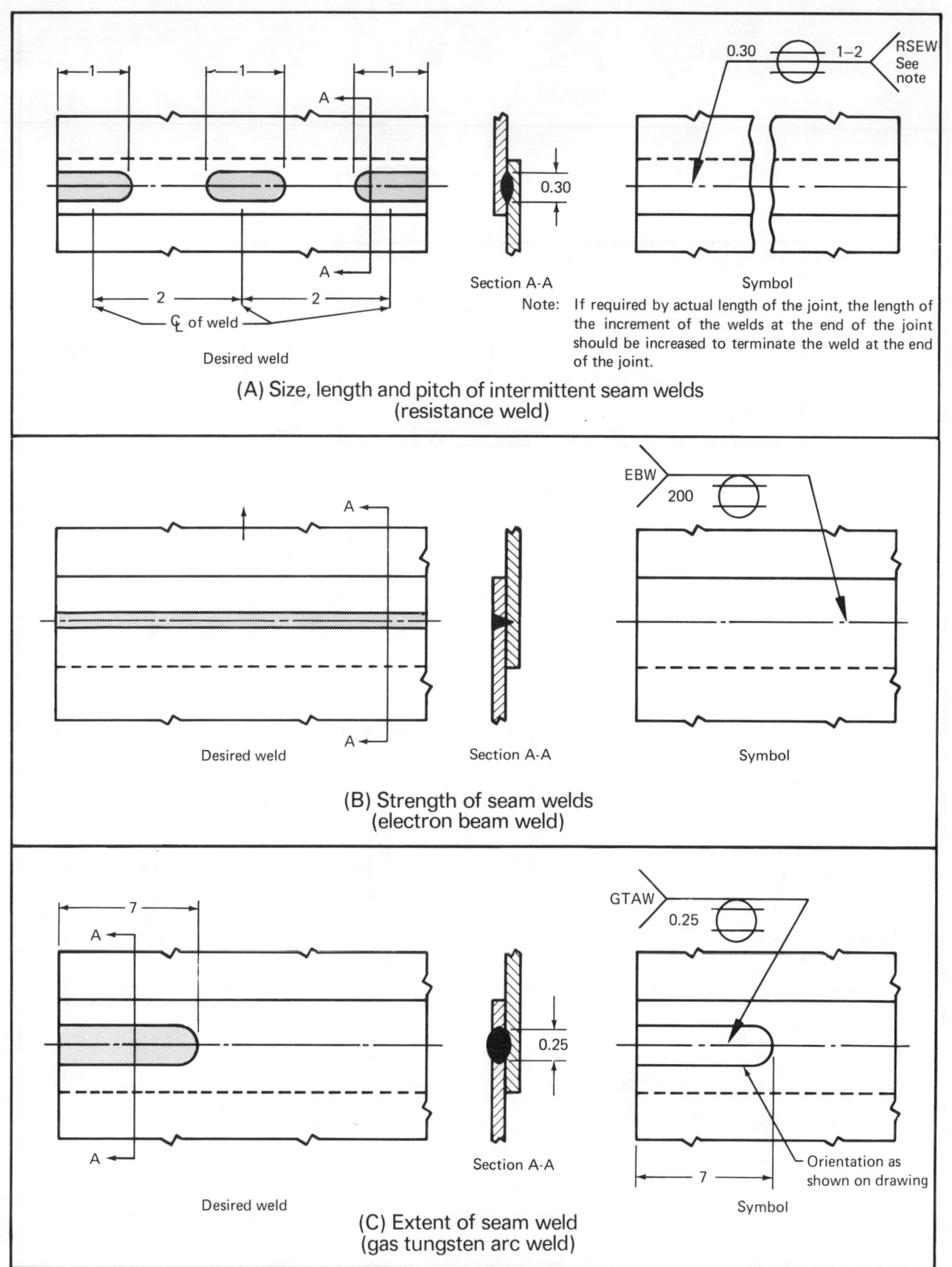

Figure 37 — Application of Dimensions to Seam Weld Symbols

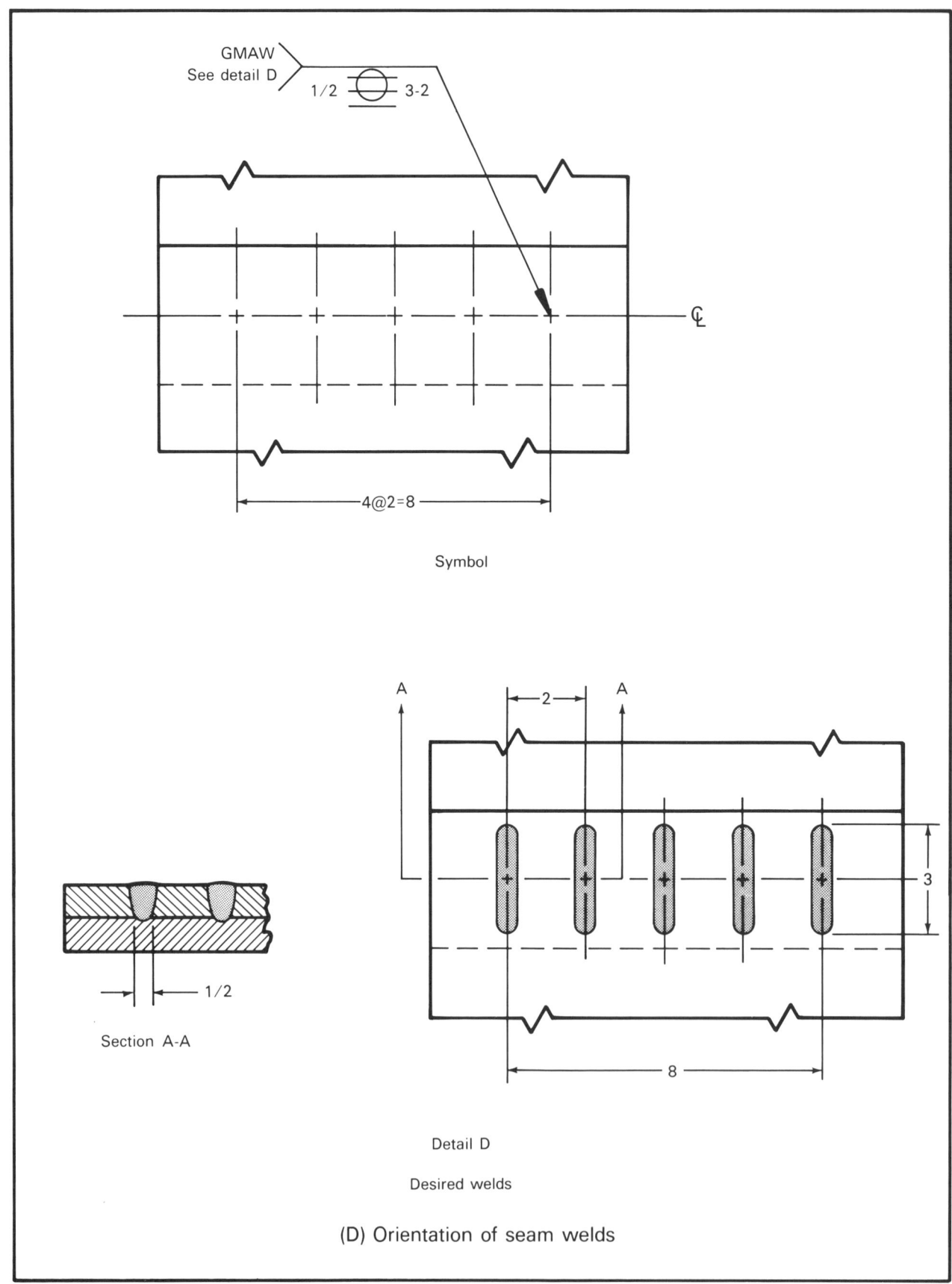

Figure 37 (continued) — Application of Dimensions to Seam Weld Symbols

direction of the welding, or less than the full length of the joint, the extent shall be dimensioned [see Figure 37(C)].

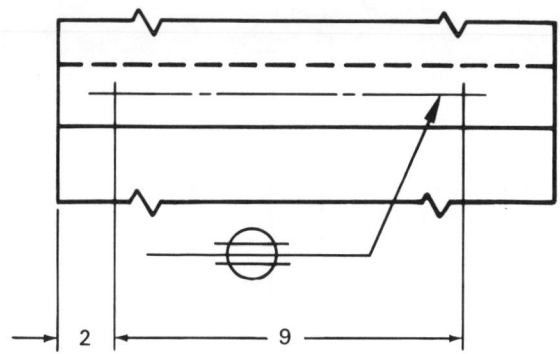

9.4 Dimensions of Intermittent Seam Welds

9.4.1 Pitch. The pitch of intermittent seam welds shall be specified as the distance between centers of the weld increments [see Figure 37(A) and (D)].

9.4.2 Dimension Location. The pitch of intermittent seam welds shall be specified to the right of the length dimension [see Figure 37 (A) and (D)].

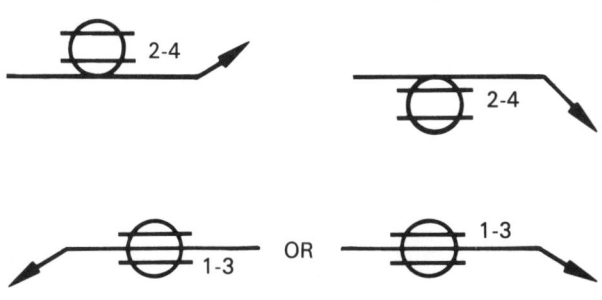

9.5 Orientation of Seam Welds

9.5.1 Intermittent Welds. Unless otherwise indicated, intermittent seam welds shall be interpreted as having length and pitch measured parallel to the weld axis [see Figure 37(A)].

9.5.2 Showing Orientation. When the orientation of seam welds is not as in 9.5.1, a detailed drawing shall be used to specify the weld orientation [see Figure 37(D)].

9.6 Contours and Finishing of Seam Welds

9.6.1 Contours Obtained by Welding. When the exposed surface of either member of a seam welded joint is to be welded with approximately flush or convex faces without postweld finishing, that surface shall be specified by adding the flush or convex contour symbol to the weld symbol.

9.6.2 Contours Obtained by Postweld Finishing. Seam welds whose faces are to be finished approximately flush or convex by postweld finishing shall be specified by adding both the appropriate contour and finishing symbol to the welding symbol (see 3.12.1). Welds that require a flat but not flush surface require an explanatory note in the tail of the symbol.

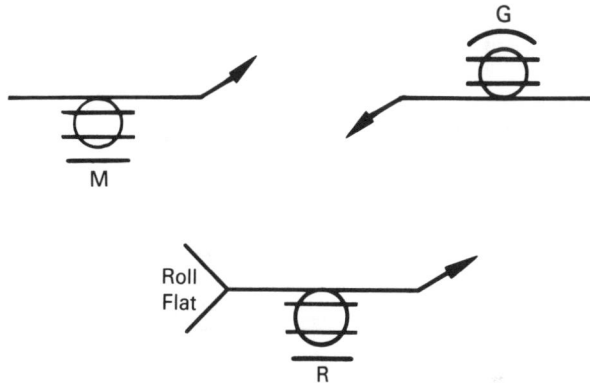

9.7 Multiple-Joint Seam Welds. When one or more members are included between the two outer members in a seam welded joint, the seam weld symbol for the two outer pieces shall be used.

10. Flange Welds

10.1 General. The following welding symbols are intended to be used for light gage metal joints where flanging of the edges is required.

10.1.1 Edge-Flange Welds. Edge-flange welds shall be specified by the edge-flange weld symbol on joints either detailed or not detailed on the drawing. (This symbol does not have both side significance) (see Figure 38).

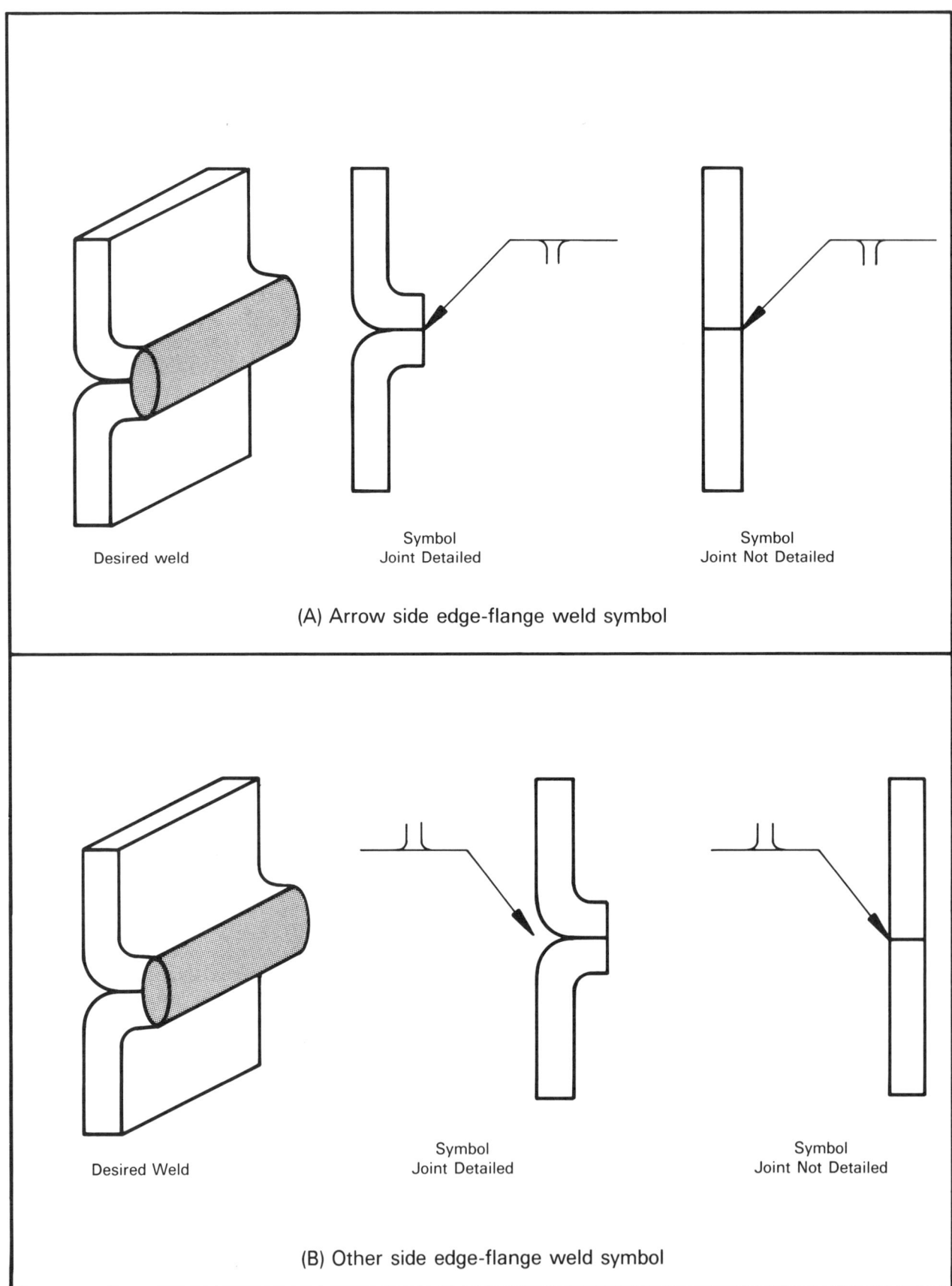

Figure 38 — Application of Edge-Flange Weld Symbol

10.1.2 Corner-Flange Welds

10.1.2.1 Joint Detailed. Corner flange welds on joints detailed on the drawing shall be specified by the corner-flange weld symbol. (This symbol does not have both side significance) (see Figure 39).

10.1.2.2 Joint not Detailed. Corner-flange welds on joints not detailed on the drawing shall be specified by the corner-flange weld symbol. A broken arrow shall point to the member to be flanged. (This symbol does not have both side significance) (see Figure 39).

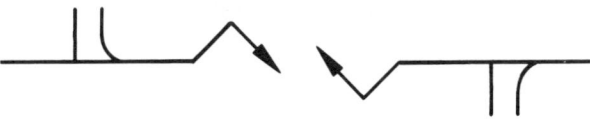

10.1.3 Edge-Flange Welds Requiring Complete Joint Penetration.
Edge-flange welds requiring complete joint penetration shall be specified by the edge-flange weld symbol with the melt-through symbol placed on the opposite side of the reference line. The same welding symbol is used for joints either detailed or not detailed on the drawing [see Figure 10(D)].

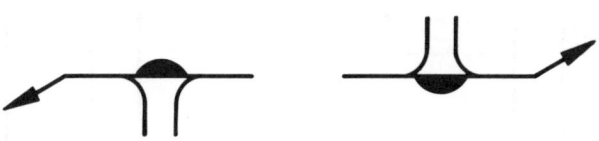

10.1.4 Corner-Flange Welds Requiring Complete Joint Penetration

10.1.4.1 Joint Detailed. Corner-flange welds requiring complete joint penetration shall be specified by the corner-flange weld symbol with the melt-through symbol placed on the opposite side of reference line [see Figure 10(E)].

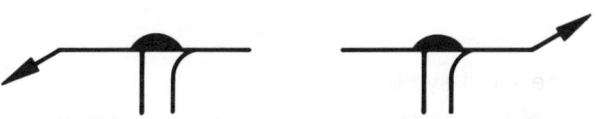

10.1.4.2 Joint not Detailed. Corner-flange welds requiring complete joint penetration shall be specified by the corner-flange weld symbol with the melt-through symbol placed on the opposite side of the reference line. A broken arrow shall point to the member to be flanged [see Figure 10(E)].

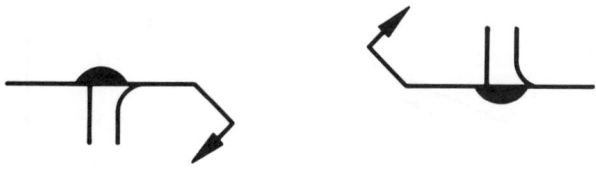

10.2 Dimensions of Flange Welds

10.2.1 Location. Dimensions of flange welds shall be placed on the same side of the reference line as the weld symbol (see Figure 40).

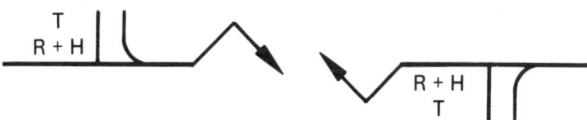

Where: T = Weld thickness
        H = Height of flange
        R = Radius of flange

10.2.2 Radius and Height. The radius and height above the point of tangency shall be indicated by specifying both the radius and the height, separated by a plus sign, and placed to the left of the weld symbol. The radius and the height shall read in that order from left to right along the reference line (see Figure 40).

10.2.3 Weld Size (Thickness). The size (thickness) of flange welds shall be specified by a dimension placed above or below the flange dimensions as applicable (see Figure 40).

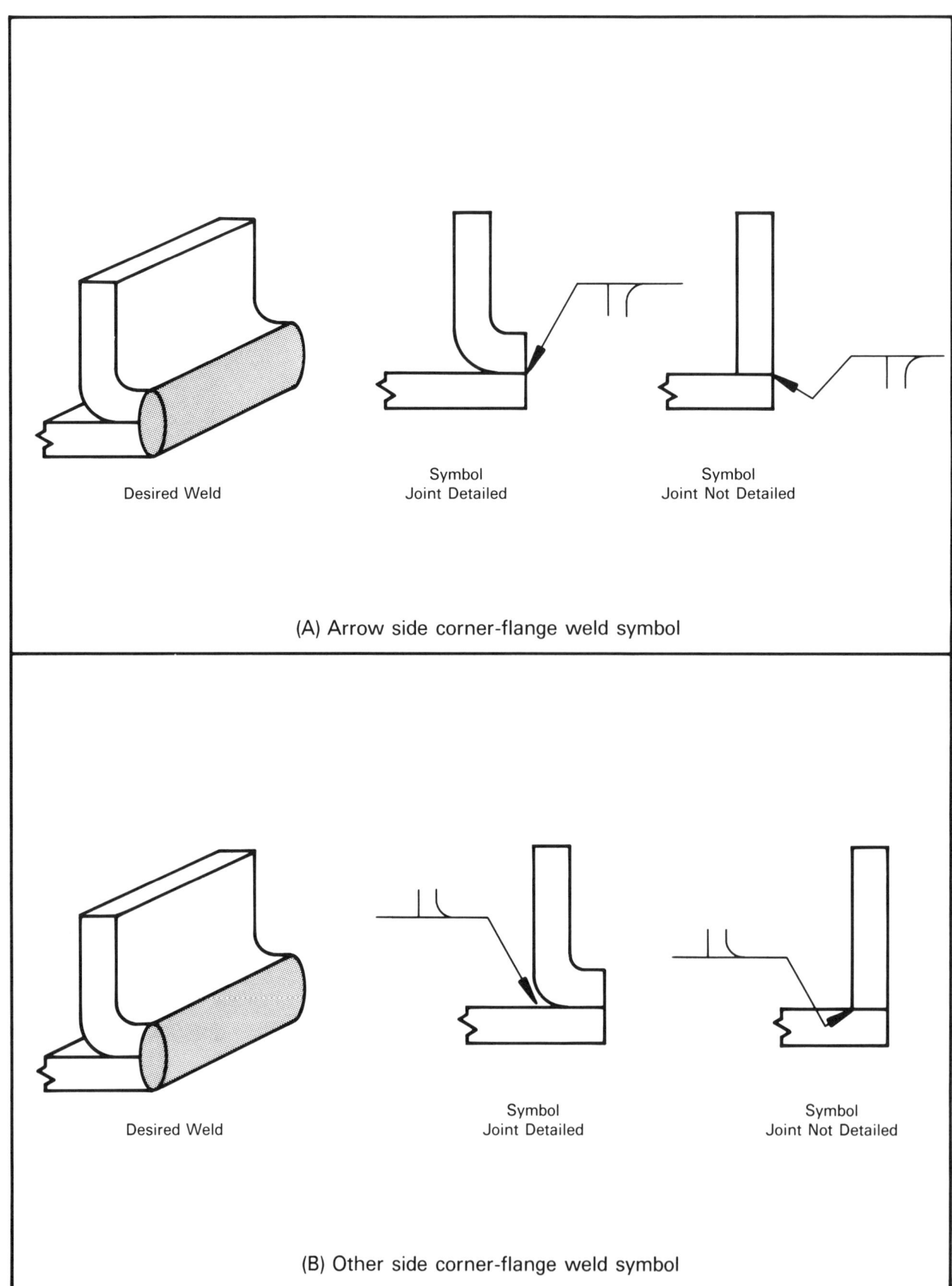

Figure 39 — Application of Corner-Flange Weld Symbol

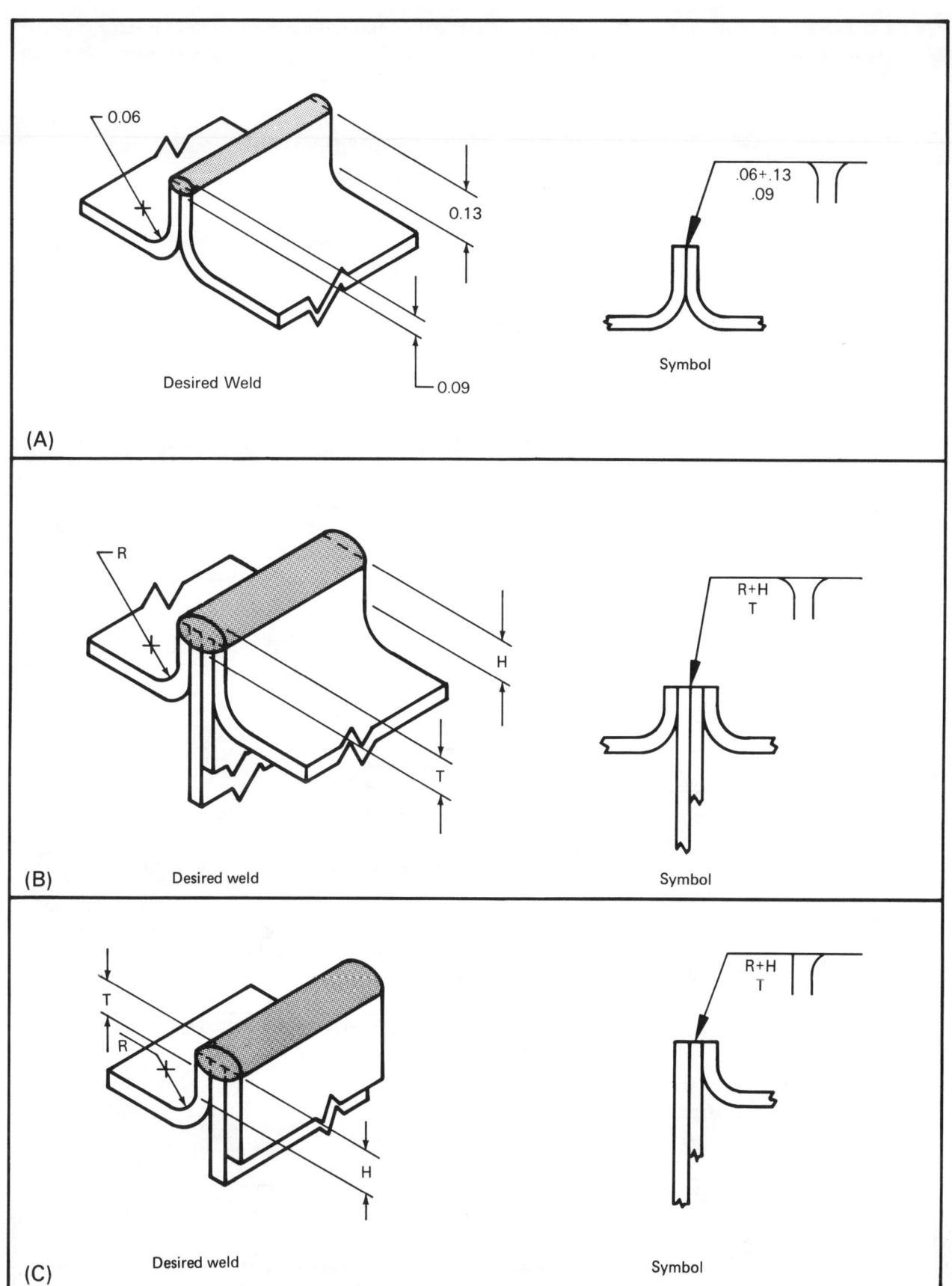

Figure 40 — Application of Edge and Corner-Flange Weld Symbols

10.2.4 Root Opening. The root opening of flange welds shall not be specified on the welding symbol. If a root opening is required, it shall be specified on the drawing.

10.3 Multiple-Joint Flange Welds. When one or more pieces are inserted between the two outer pieces in a flange welded joint, the applicable symbol for the two outer pieces shall be used [see Figure 40(B) and (C)].

Stud Welds

11.1 Side Significance. The stud weld symbol does not indicate the welding of a joint in the ordinary sense, and, therefore, has no arrow- or other-side significance. The symbol shall be placed below the reference line and the arrow shall point clearly to the surface to which the stud is to be welded.

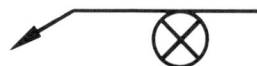

11.2 Stud Size. The required diameter of the stud shall be specified to the left side of the weld symbol (see Figure 41).

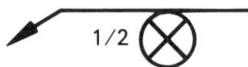

11.3 Spacing of Stud Welds. The pitch (center-to-center distance) of stud welds in a straight line, if specified, shall be to the right of the weld symbol (see Figure 41).

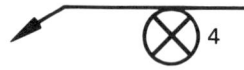

The spacing of stud welds in any configuration other than a straight line shall be dimensioned on the drawing.

11.4 Number of Stud Welds. The number of stud welds shall be specified in parentheses below the stud weld symbol (see Figure 41).

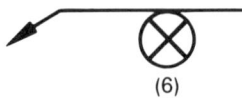

11.5 Dimension Locations. Dimensions shall be placed on the same side of the reference line as the stud weld symbol (see Figure 41).

11.6 Location of First and Last Stud Welds. The location of the first and last stud weld in each single line shall be specified on the drawing [see Figure 41(B)].

12. Surfacing Welds

12.1 Use of Surfacing Weld Symbol

12.1.1 Symbol Application. Surfacing, whether by single- or multiple-pass welds, shall be specified by the surfacing weld symbol (see Figure 42).

12.1.2 Arrow-Side Significance. The surfacing weld symbol does not indicate the welding of a joint, and, therefore, has no arrow- or other-side significance. the symbol shall be placed below the reference line, and the arrow shall point clearly to the surface on which the surfacing weld is to be made (see Figure 42).

12.1.3 Dimension Location. Dimensions used in conjunction with the surfacing weld symbol shall be placed on the same side of the reference line as the weld symbol (see Figure 42).

12.2 Size (Thickness) of Surfacing Welds

12.2.1 Minimum Thickness. The size (thickness) of a surfacing weld shall be specified by placing the dimension of the required thickness to the left of the weld symbol [see Figure 42(A) and (C)]. The direction of welding may be specified by a note in the tail of the welding symbol or indicated on the drawing.

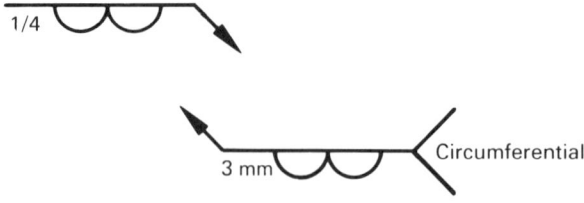

12.2.2 Multiple Layer. Multiple layer surfacing welds may be specified by using multiple reference lines with

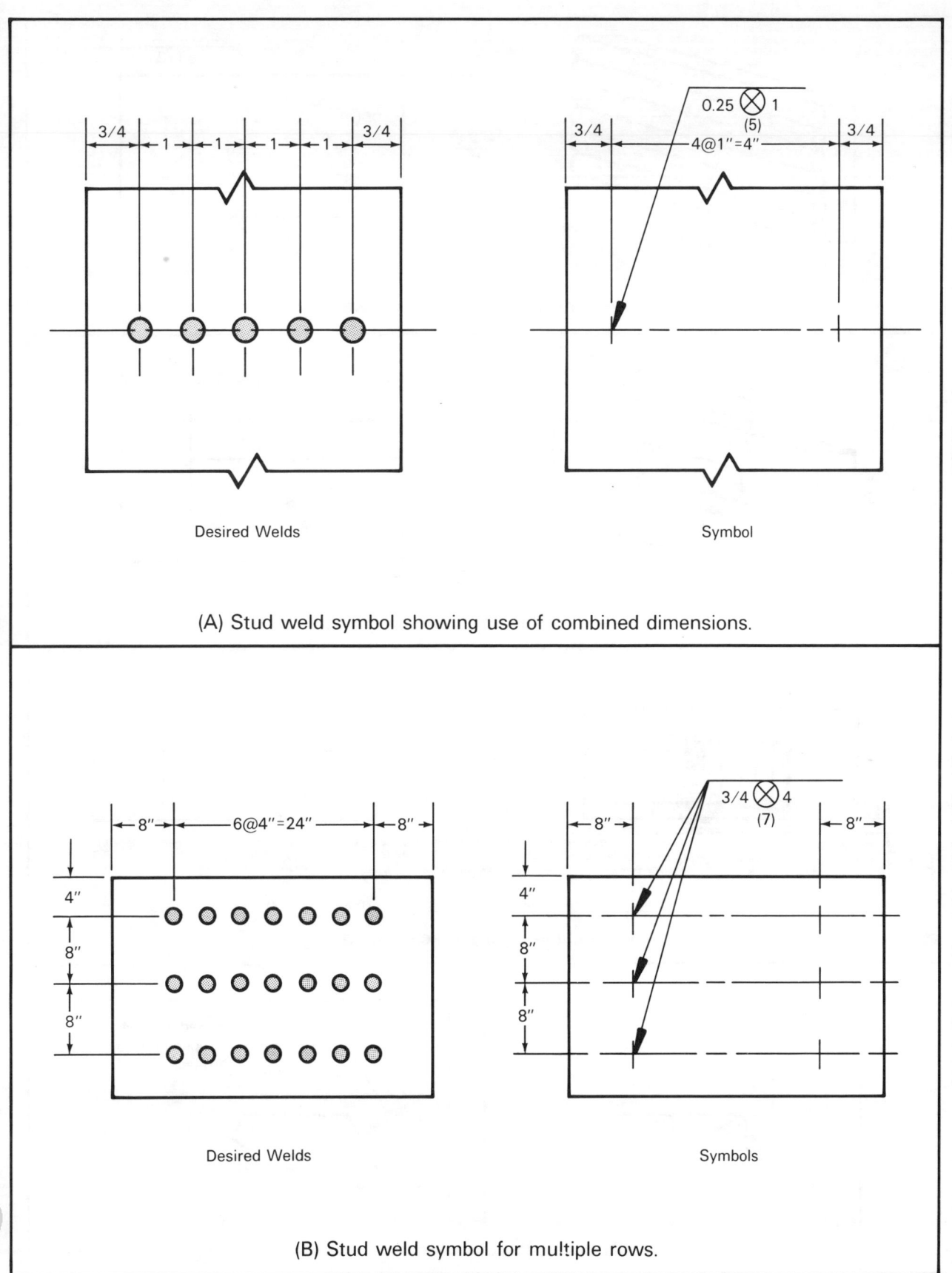

Figure 41 — Application of Stud Weld Symbols

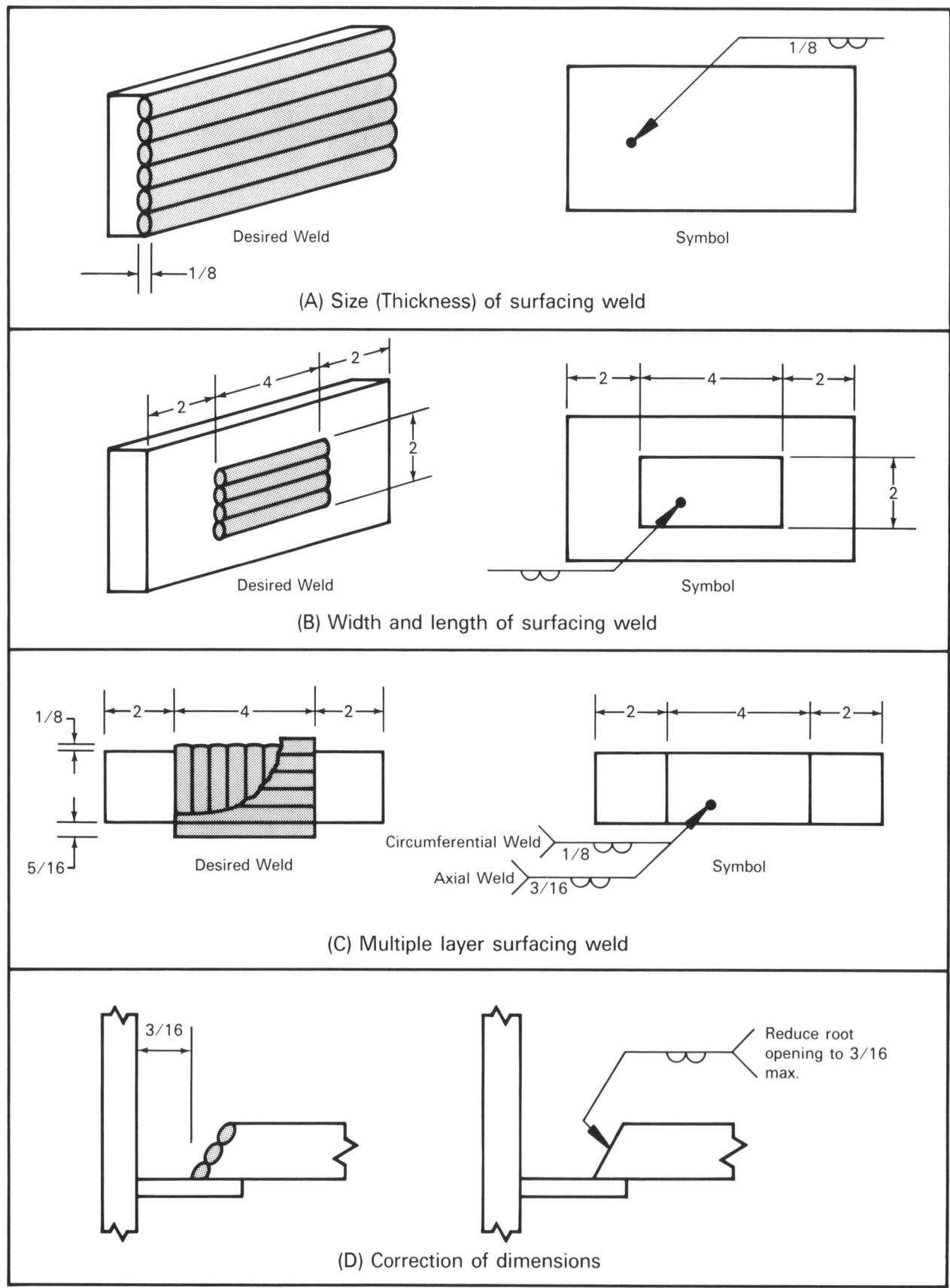

Figure 42 — Application of Surfacing Weld Symbol

the required size (thickness) of each layer placed to the left of the weld symbols. The direction of welding may be specified by an appropriate note in the tail of the welding symbol or indicated on the drawing [see Figure 42(C)].

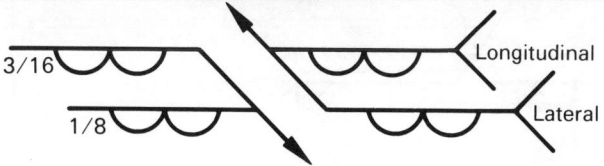

12.2.3 Unspecified Size (Thickness). When no specific thickness of a surfacing weld is required, the size dimension need not be included in the welding symbol [see Figure 42(B)].

12.3 Extent, Location, and Orientation of Surfacing Welds

12.3.1 Entire Area. No dimension other than size (thickness) is necessary to specify surfacing of the entire area of a plane or curved surface [see Figure 42(A)].

12.3.2 Portion of Area. When only a portion of a surface is to receive a surfacing weld, the extent, location, and orientation shall be shown on the drawing [see Figure 42(B) and (C)].

12.4 Surfacing of a Weld. Multiple reference lines may be used to specify a surfacing weld on the surface of a previously made weld (see 3.6).

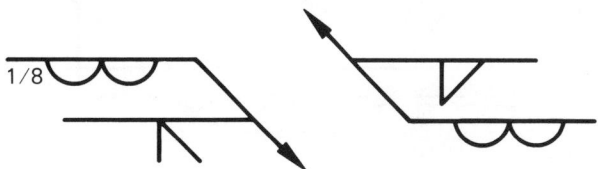

12.5 Surfacing to Adjust Dimensions. The surfacing weld symbol may be used to specify a surfacing weld to correct assembly problems such as reducing excessive root openings [see Figure 42(D)].

Part B
Brazing Symbols

13. Brazed Joints

If no special joint preparation other than cleaning is required, only the arrow and reference line need be used with the brazing process indicated in the tail [see Figure 43(A)]. Applications of conventional welding symbols to brazed joints are illustrated in Figure 43(B) through (H). Figure 43(C), (D), (E), (G), and (H) shows how joint clearances can be indicated. All symbols used for welding may also be used for brazing, where suitable.

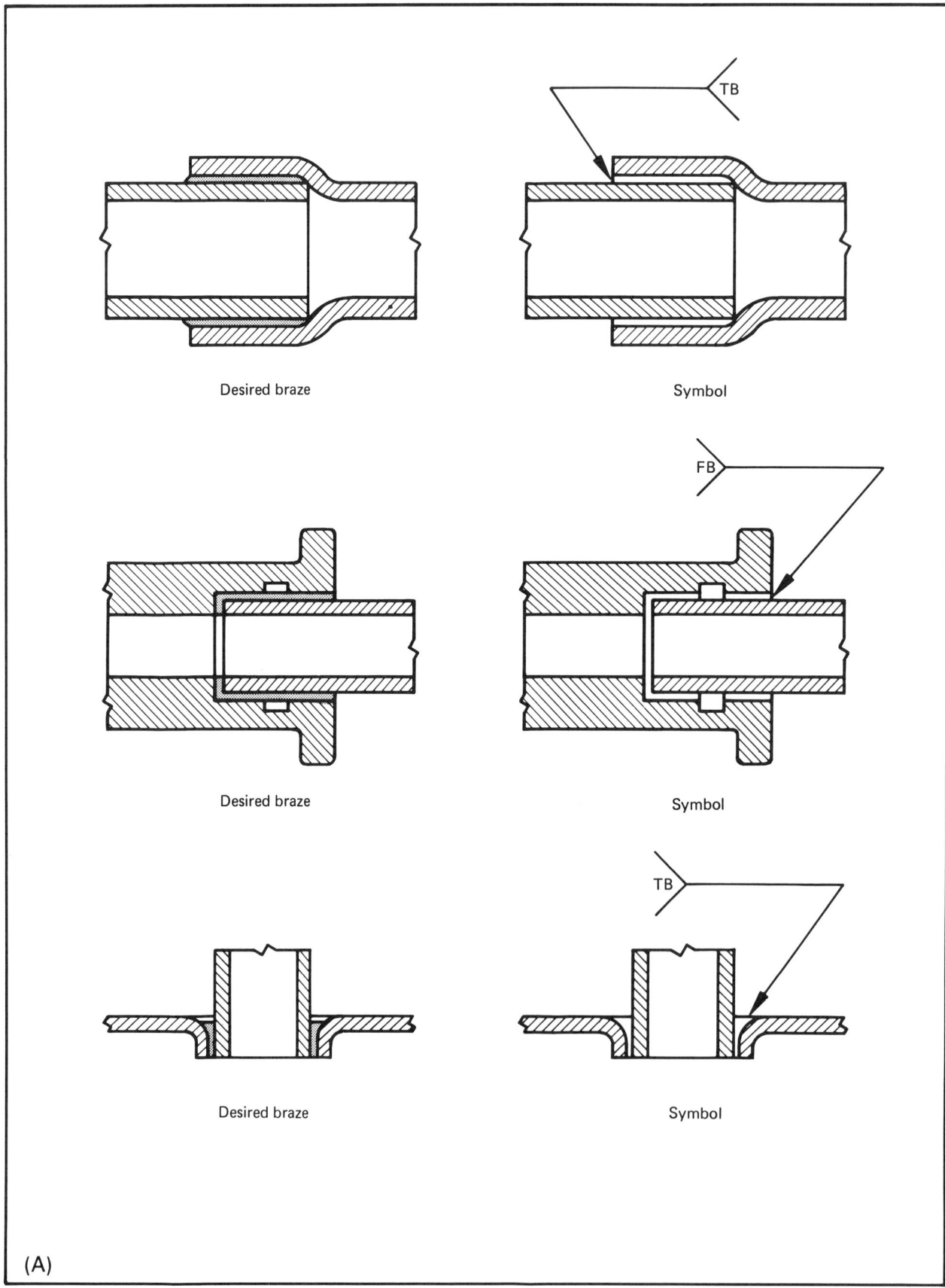

Figure 43 — Application of Brazing Symbols

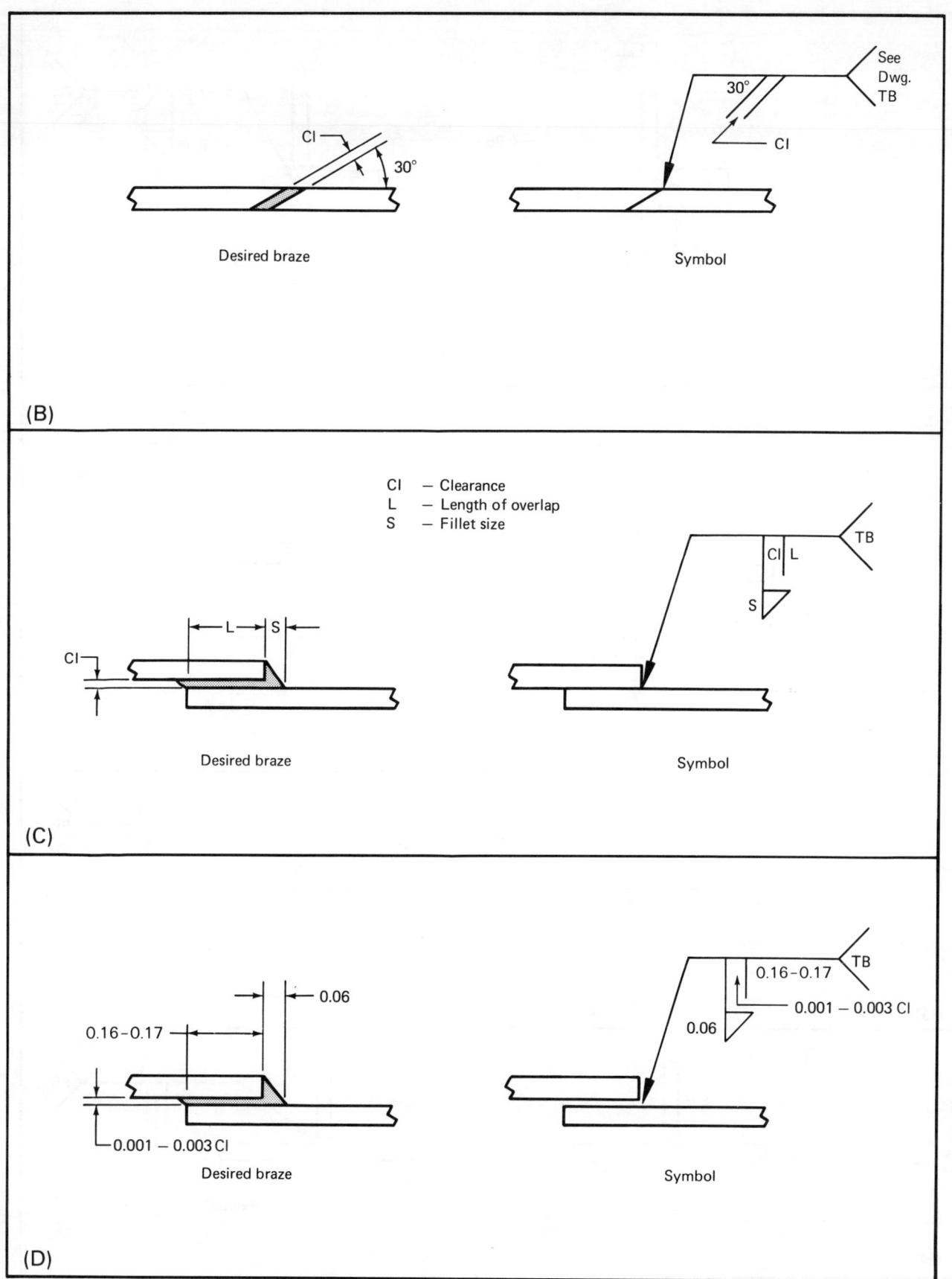

Figure 43 (continued) — Application of Brazing Symbols

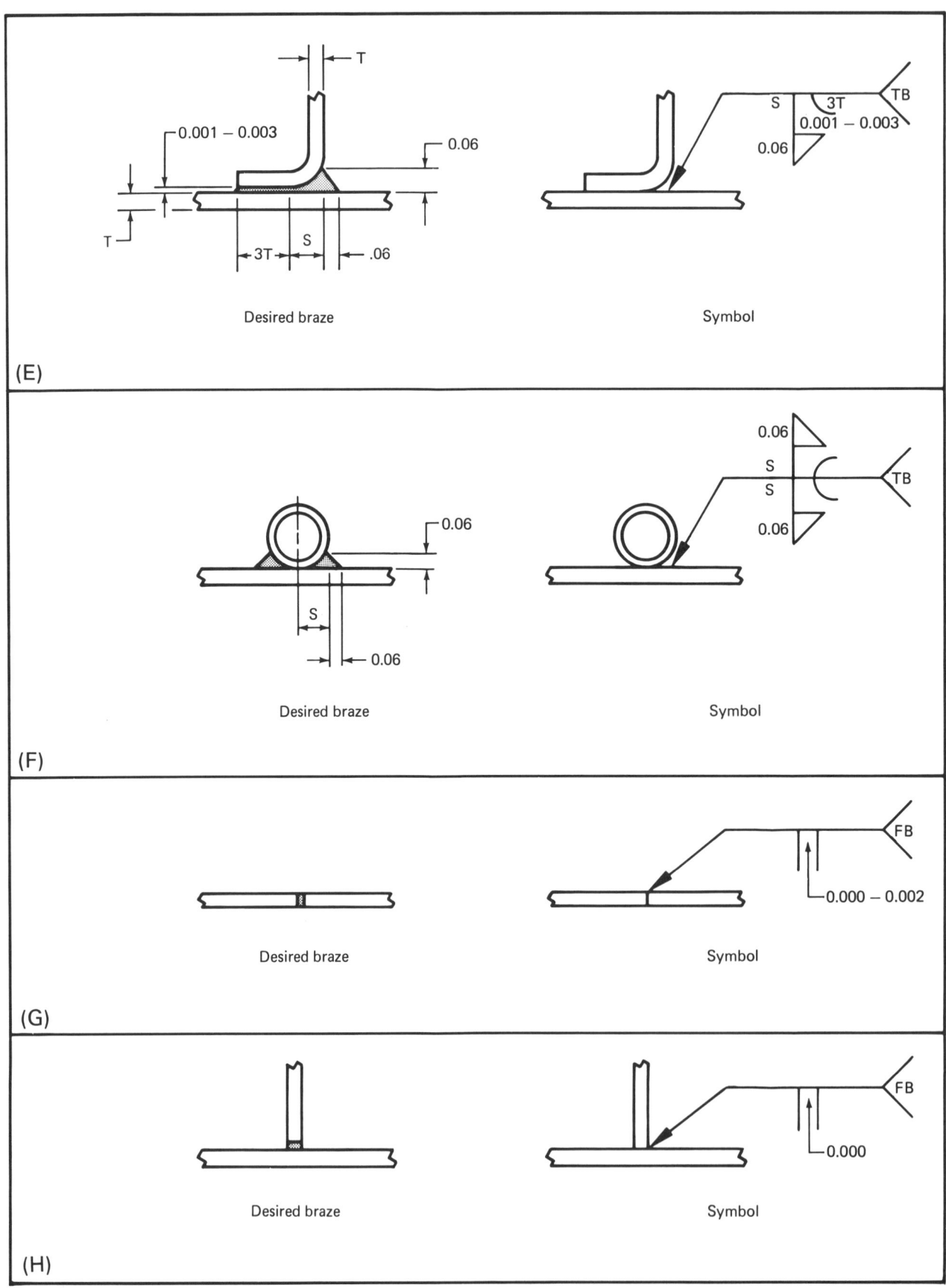

Figure 43 (continued) — Application of Brazing Symbols

Part C
Nondestructive Examination Symbols

14. Elements of the Nondestructive Examination Symbol

The examination symbol consists of the following elements:

(1) Reference line (drawn horizontally)
(2) Arrow
(3) Examination method letter designations
(4) Dimensions, areas, and number of examinations
(5) Supplementary symbols
(6) Tail
(7) Specifications, codes, or other references

Only those elements required to specify the examination requirements need be included in a nondestructive examination symbol.

14.1 Examination Method Letter Designations. Nondestructive examination methods shall be specified by use of the letter designation shown below.

Examination Method	Letter Designation
Acoustic emission	AET
Electromagnetic	ET
Leak	LT
Magnetic particle	MT
Neutron radiographic	NRT
Penetrant	PT
Proof	PRT
Radiographic	RT
Ultrasonic	UT
Visual	VT

14.2 Supplementary Symbols. Supplementary symbols to be used in nondestructive examination symbols shall be as follows:

Examine All Around	Field Examination	Radiation Direction
⌀	⚑	✵

14.3 Standard Location of Elements of a Nondestructive Examination Symbol. The elements of a nondestructive examination symbol shall have standard locations with respect to each other as shown in Figure 44.

15. General Provisions

15.1 Location Significance of Arrow. The arrow shall connect the reference line to the part to be examined. The side of the part to which the arrow points shall be considered the arrow side of the part. The side opposite the arrow side of the part shall be considered the other side.

15.2 Location of Letter Designations

15.2.1 Location on Arrow Side. Examinations to be made on the arrow side of the part shall be specified by placing the letter designation for the selected examination method below the reference line.

15.2.2 Location on the Other Side. Examinations to be made on the other side of the part shall be specified by placing the letter designation for the selected examination method above the reference line.

15.2.3 Location on Both Sides. Examinations to be made on both sides of the part shall be specified by placing the letter designation for the selected examination method on both sides of the reference line.

15.2.4 Location Centered on Reference Line. When the letter designation has no arrow or other side signifi-

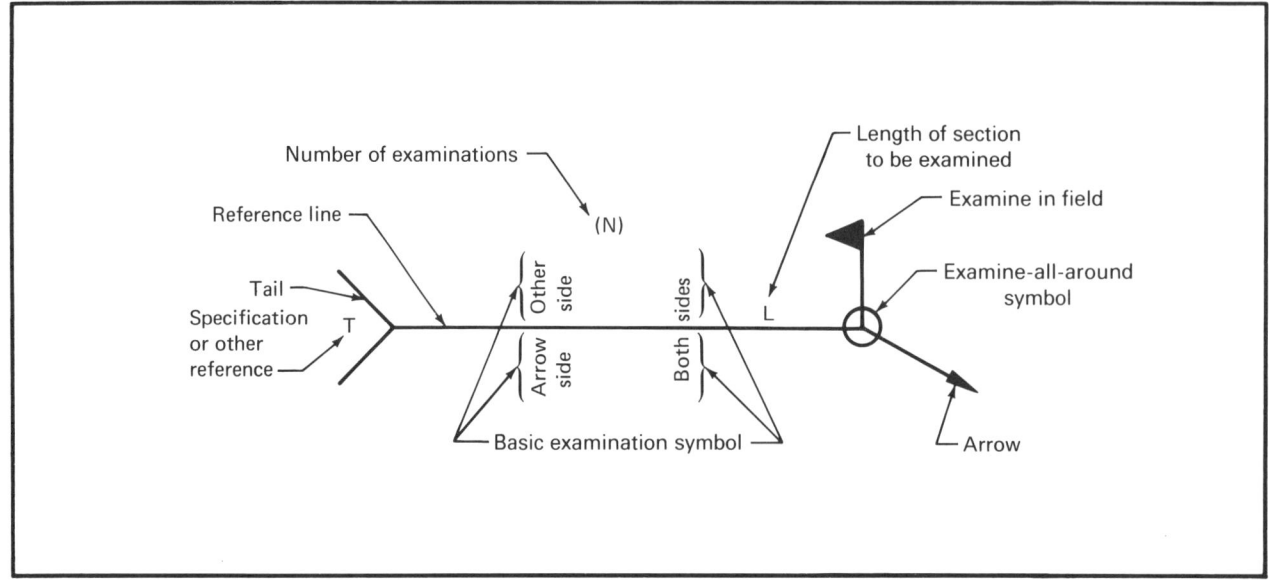

Figure 44 — Standard Location of Elements

cance, the letter designation shall be centered on the reference line.

15.2.5 Examination Combinations. More than one examination method may be specified for the same part by placing the combined letter designations of the selected examination methods in the appropriate positions relative to the reference line. Letter designations for two or more examination methods, to be placed on the same side of the reference line or centered on the reference line, shall be separated by a plus sign.

15.2.6 Welding and NDT Symbols. Nondestructive testing symbols and welding symbols may be combined.

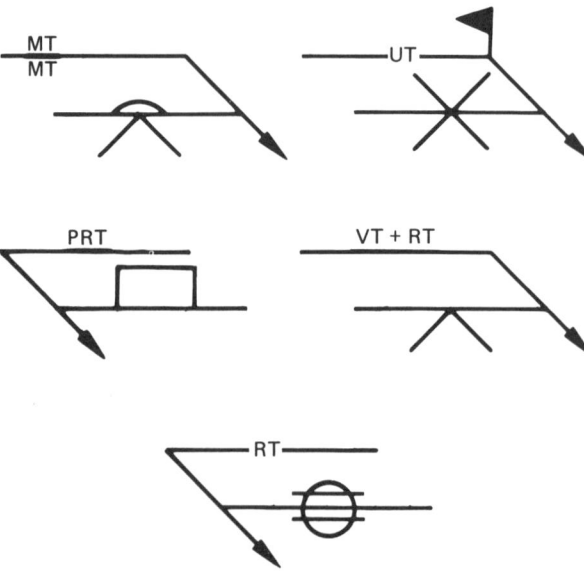

15.3 U.S. Customary and Metric Units. When it is required to specify dimensions with nondestructive examination symbols, the same system of units that is standard for the drawing shall be used. Dual units shall not be used on nondestructive examination symbols. If it is required to include conversions from metric to U.S. customary, or vice versa, a table of conversions may be included on the drawing. For guidance in drafting standards, reference is made to the ANSI Y14, *Drafting Manual*. For guidance on the use of metric (SI) units, reference is made to AWS A1.1, *Metric Practice Guide for the Welding Industry*.

16. Supplementary Symbols

16.1 Examine-All-Around. Examinations required all around a weld, joint, or part shall be specified by placing the examine-all-around symbol at the junction of the arrow and reference lines.

16.2 Field Examinations. Examinations required to be conducted in the field (not in a shop or at the place of initial construction) shall be specified by placing the field examination symbol at the junction of the arrow and reference lines.

16.3 Radiation Direction. The direction of penetrating radiation may be specified by use of the radiation direction symbol drawn at the required angle on the drawing and the angle indicated in degrees if necessary to ensure clarity.

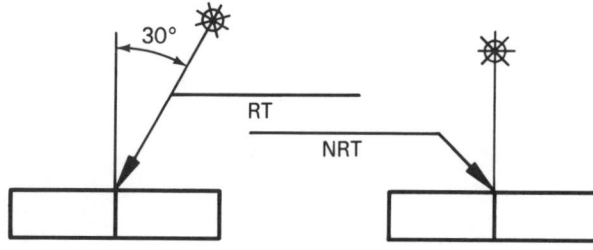

17. Specifications, Codes, and References

Information, applicable to the examination specified, which is not otherwise provided, may be placed in the tail of the nondestructive examination symbol.

18. Location, Orientation, and Extent of Nondestructive Examination

18.1 Specifying Length of Section to be Examined

18.1.1 Length Shown. To specify examination of welds or parts where only the length of a section need be considered, the length dimension shall be placed to the right of the letter designation.

18.1.2 Location Shown. To specify the exact location of a section to be examined, as well as the length, dimension lines shall be used.

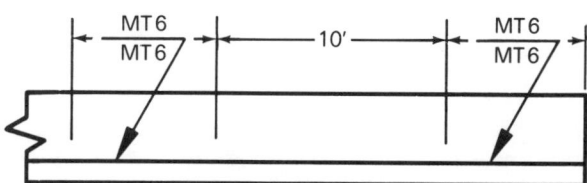

18.1.3 Full Length Examination. When the full length of a part is to be examined, no length dimension need be included in the nondestructive examination symbol.

18.1.4 Partial Examination. When less than one hundred percent of the length of a weld or part is to be examined with locations to be determined by a specified procedure, the length to be examined is specified by placing the appropriate percentage to the right of the letter designation. The selected procedure may be specified by reference in the tail.

18.2 Number of Examinations. To specify a number of examinations to be conducted on a joint or part at random locations, the number of required examinations shall be placed in parentheses either above or below the letter designation away from the reference line.

18.3 Examination of Areas. Nondestructive examination of areas shall be specified by one of the following methods:

18.3.1 Plane Areas. To specify nondestructive examination of an area represented as a plane on the drawing, the area to be examined shall be enclosed by straight, broken lines with a circle at each change in direction. The letter designations for the nondestructive examination required shall be used in connection with these lines as shown below. When necessary, these enclosures shall be located by coordinate dimensions.

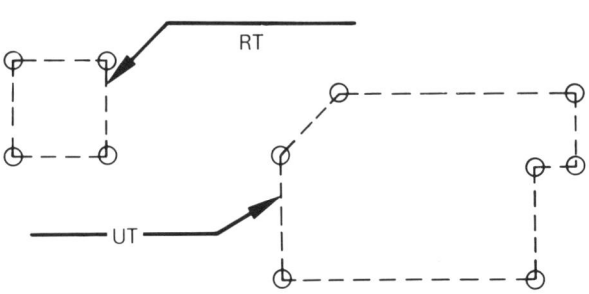

18.3.2 Areas of Revolution. For nondestructive examination of areas of revolution, the area shall be specified by using the examine-all-around symbol and appropriate dimensions. The following illustration indicates:

18.3.2.1 Upper right hand symbol. Magnetic particle examination of the bore of the flange for a distance of three inches from the right hand face, all the way around.

18.3.2.2 Lower left hand symbol. Radiographic examination of an area of revolution where dimensions were not available on the drawing.

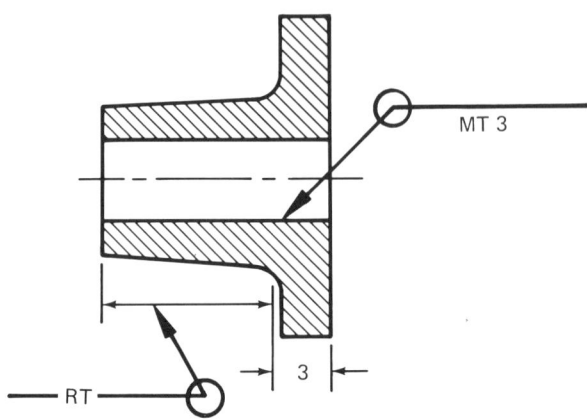

The symbol below specifies an area of revolution subject to an internal proof examination and an external eddy current examination. Since no dimensions are given, the entire length is to be examined.

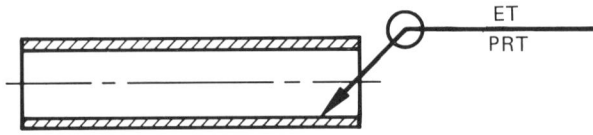

18.3.3 Acoustic Emission. Acoustic emission is generally applied to all or a large portion of a component, such as a pressure vessel or pipe. The symbol indicates application of AET to the component without specific reference to location of sensors.

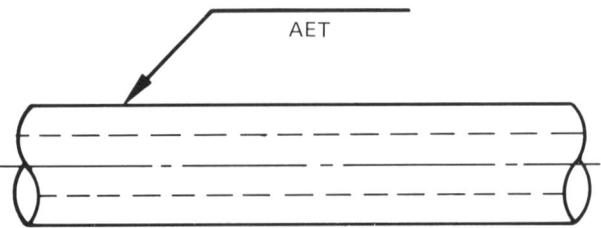

Table 1
Designation of Welding and Allied Processes by Letters

Welding and allied processes	Letter designation	Welding and allied processes	Letter designation
adhesive bonding	ABD	induction	RSEW-I
arc welding	AW	resistance spot welding	RSW
atomic hydrogen welding	AHW	upset welding	UW
bare metel arc welding	BMAW	high frequency	UW-HF
carbon arc welding	CAW	induction	UW-I
gas	CAW-G	soldering	S
shielded	CAW-S	dip soldering	DS
twin	CAW-T	furnace soldering	FS
electrogas	EGW	induction soldering	IS
flux cored arc welding	FCAW	infrared soldering	IRS
gas metal arc welding	GMAW	iron soldering	INS
pulsed arc	GMAW-P	resistance soldering	RS
short circuiting arc	GMAW-S	torch soldering	TS
gas tungsten arc welding	GTAW	wave soldering	WS
pulsed arc	GTAW-P	solid-state welding	SSW
plasma arc welding	PAW	coextrusion welding	CEW
shielded metal arc welding	SMAW	cold welding	CW
stud arc welding	SW	diffusion welding	DFW
submerged arc welding	SAW	explosion welding	EXW
series	SAW-S	forge welding	FOW
brazing	B	friction welding	FRW
arc brazing	AB	hot pressure welding	HPW
block brazing	BB	roll welding	ROW
carbon arc brazing	CAB	ultrasonic welding	USW
diffusion brazing	DFB	thermal cutting	TC
dip brazing	DB	arc cutting	AC
flow brazing	FLB	air carbon arc cutting	AAC
furnace brazing	FB	carbon arc cutting	CAC
induction brazing	IB	gas metal arc cutting	GMAC
infrared brazing	IRB	gas tungsten arc cutting	GTAC
resistance brazing	RB	metal arc cutting	MAC
torch brazing	TB	plasma arc cutting	PAC
other welding processes		shielded metal arc cutting	SMAC
electron beam welding	EBW	electron beam cutting	EBC
high vacuum	EBW-HV	laser beam cutting	LBC
medium vacuum	EBW-MV	air	LBC-A
nonvacuum	EBW-NV	evaporative	LBC-EV
electroslag welding	ESW	inert gas	LBC-IG
flow welding	FLOW	oxygen	LBC-O
induction welding	IW	oxygen cutting	OC
laser beam welding	LBW	chemical flux cutting	FOC
thermit welding	TW	metal powder cutting	POC
oxyfuel gas welding	OFW	oxyfuel gas cutting	OFC
air acetylene welding	AAW	oxyacetylene cutting	OFC-A
oxyacetylene welding	OAW	oxyhydrogen cutting	OFC-H
oxyhydrogen welding	OHW	oxynatural gas cutting	OFC-N
pressure gas welding	PGW	oxypropane cutting	OFC-P
resistance welding	RW	oxygen arc cutting	AOC
flash welding	FW	oxygen lance cutting	LOC
percussion welding	PEW	thermal spraying	THSP
projection welding	PW	arc spraying	ASP
resistance seam welding	RSEW	flame spraying	FLSP
high frequency	RSEW-HF	plasma spraying	PSP

Table 2
Alphabetical Cross Reference to Table 1

Letter designation	Welding and allied processes	Letter designation	Welding and allied processes
AAC	air carbon arc cutting	GTAW	gas tungsten arc welding
AAW	air acetylene welding	GTAW-P	gas tungsten arc welding—plused arc
ABD	adhesive bonding		
AB	arc brazing	HPW	hot pressure welding
AC	arc cutting	IB	induction brazing
AHW	atomic hydrogen welding	INS	iron soldering
AOC	oxygen arc cutting	IRB	infrared brazing
ASP	arc spraying	IRS	infrared soldering
AW	arc welding	IS	induction soldering
B	brazing	IW	induction welding
BB	block brazing	LBC	laser beam cutting
BMAW	bare metal arc welding	LBC-A	laser beam cutting—air
CAB	carbon arc brazing	LBC-EV	laser beam cutting—evaporative
CAC	carbon arc cutting		
CAW	carbon arc welding	LBC-IG	laser beam cutting—inert gas
CAW-G	gas carbon arc welding	LBC-O	laser beam cutting—oxygen
CAW-S	shielded carbon arc welding	LBW	laser beam welding
CAW-T	twin carbon arc welding	LOC	oxygen lance cutting
CEW	coextrusion welding	MAC	metal arc cutting
CW	cold welding	OAW	oxyacetylene welding
DB	dip brazing	OC	oxygen cutting
DFB	diffusion brazing	OFC	oxyfuel gas cutting
DFW	diffusion welding	OFC-A	oxyacetylene cutting
DS	dip soldering	OFC-H	oxyhydrogen cutting
EBC	electron beam cutting	OFC-N	oxynatural gas cutting
EBW	electron beam welding	OFC-P	oxypropane cutting
EBW-HV	electron beam welding—high vacuum	OFW	oxyfuel gas welding
		OHW	oxyhydrogen welding
EBW-MV	electron beam welding—medium vacuum	PAC	plasma arc cutting
		PAW	plasma arc welding
EBW-NV	electron beam welding—nonvacuum	PEW	percussion welding
		PGW	pressure gas welding
EGW	electrogas welding	POC	metal powder cutting
ESW	electroslag welding	PSP	plasma spraying
EXW	explosion welding	PW	projection welding
FB	furnace brazing	RB	resistance brazing
FCAW	flux cored arc welding	RS	resistance soldering
FLB	flow brazing	RSEW	resistance seam welding
FLOW	flow welding	RSEW-HF	resistance seam welding—high frequency
FLSP	flame spraying		
FOC	chemical flux cutting	RSEW-I	resistance seam welding—induction
FOW	forge welding		
FRW	friction welding	RSW	resistance spot welding
FS	furnace soldering	ROW	roll welding
FW	flash welding	RW	resistance welding
GMAC	gas metal arc cutting	S	soldering
GMAW	gas metal arc welding	SAW	submerged arc welding
GMAW-P	gas metal arc welding—pulsed arc	SAW-S	series submerged arc welding
		SMAC	shielded metal arc cutting
GMAW-S	gas metal arc welding—short circuiting arc	SMAW	shielded metal arc welding
		SSW	solid-state welding
GTAC	gas tungsten arc cutting	SW	stud arc welding

Table 2 (cont.)
Alphabetical Cross Reference to Table 1

Letter designation	Welding and allied processes	Letter designation	Welding and allied processes
TB	torch brazing	UW	upset welding
TC	thermal cutting	UW-HF	upset welding—high frequency
THSP	thermal spraying		
TS	torch soldering	UW-I	upset welding—induction
TW	thermit welding	WS	wave soldering
USW	ultrasonic welding		

Table 3
Suffixes for Optional Use in Applying Welding and Allied Processes

Automatic	AU	Manual	MA
Machine	ME	Semiautomatic	SA

Appendix A
Design of Standard Symbols
(Dimensions in inches)

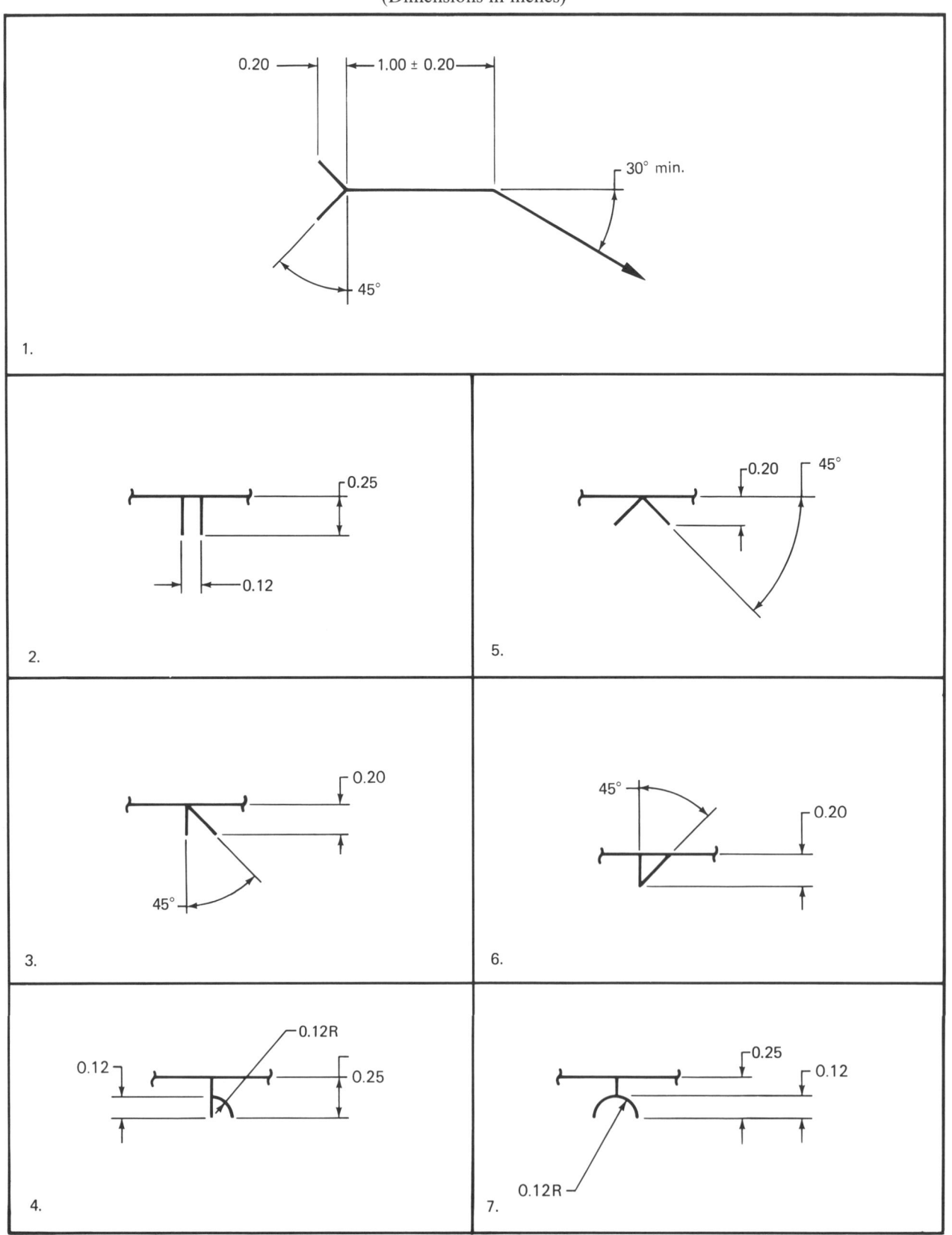

Notes: 1. Unless otherwise specified, tolerances shall be ± .04 or ± 1° as applicable.
2. All radii are minimum dimensions.

Appendix AM
Design of Standard Symbols
(Dimensions in millimeters)

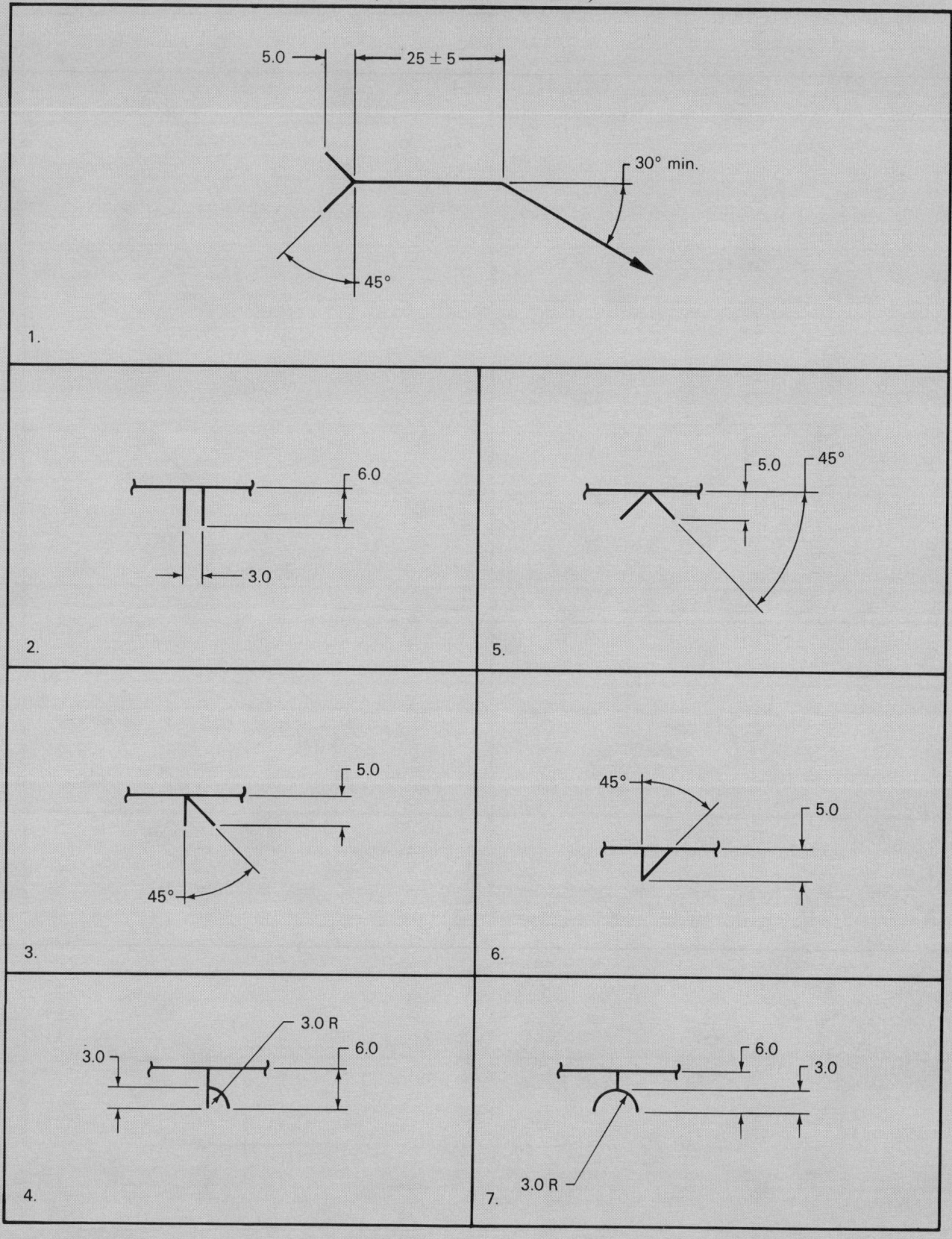

Notes: 1. Unless otherwise specified, tolerances shall be ± 1 or ± 1° as applicable.
2. All radii are minimum dimensions.

Appendix A (cont.)
Design of Standard Symbols
(Dimensions in inches)

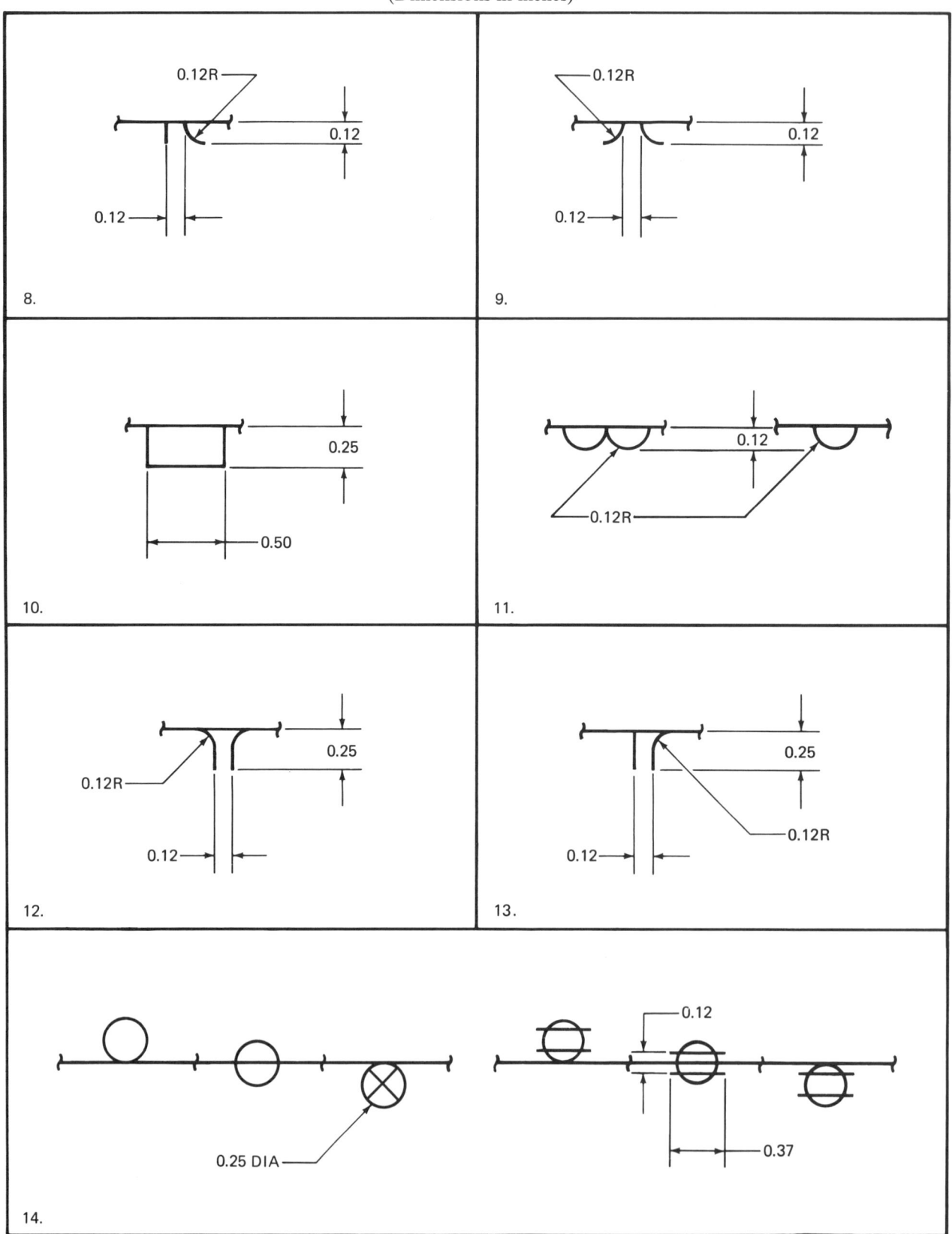

Notes: 1. Unless otherwise specified, tolerance shall be ± .04 or ± 1° as applicable.
2. All radii are minimum dimensions.

Appendix AM (cont.)
Design of Standard Symbols
(Dimensions in millimeters)

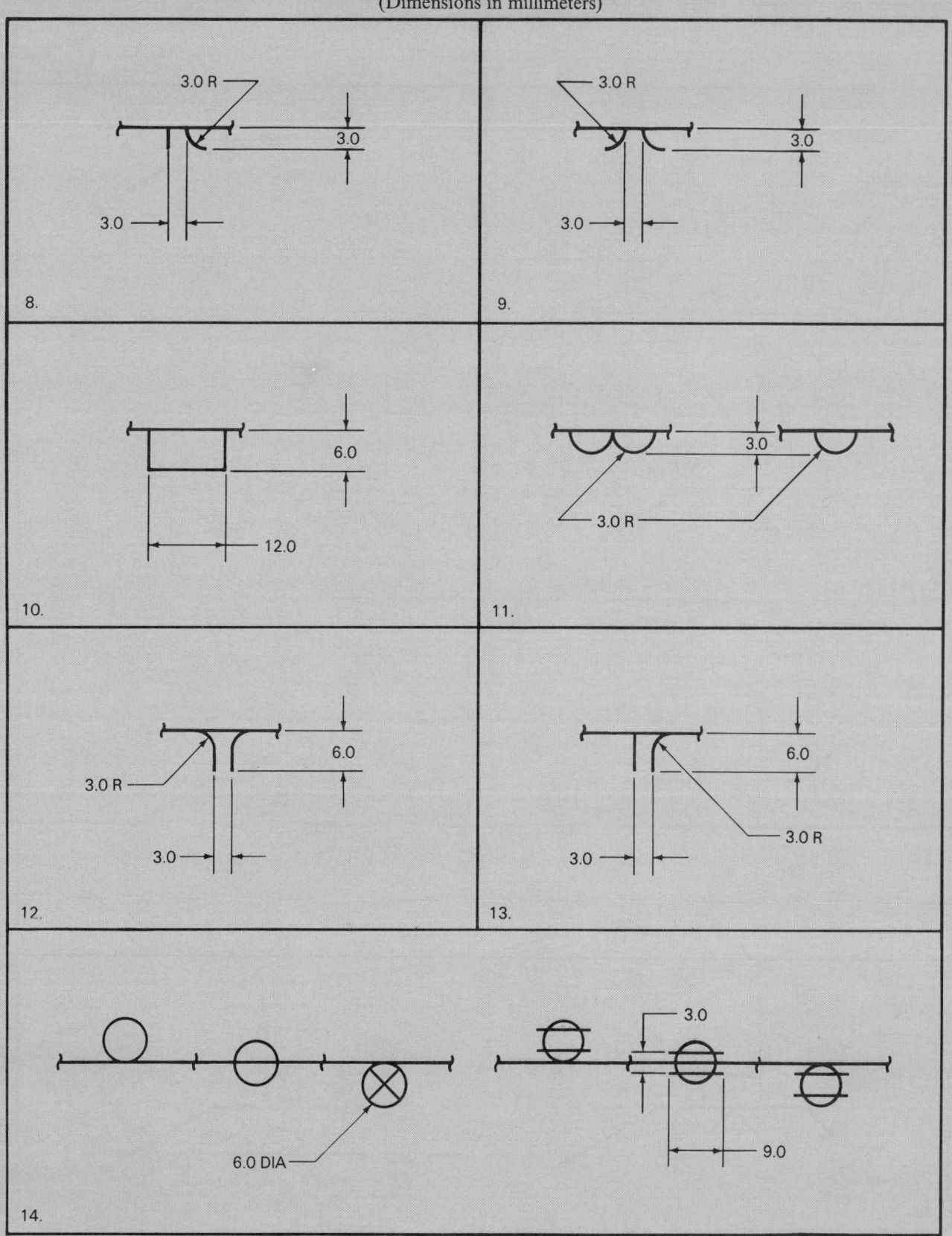

Notes: 1. Unless otherwise specified, tolerances shall be ± 1 or ± 1° as applicable.
2. All radii are minimum dimensions.

Appendix A (cont.)
Design of Standard Symbols
(Dimensions in inches)

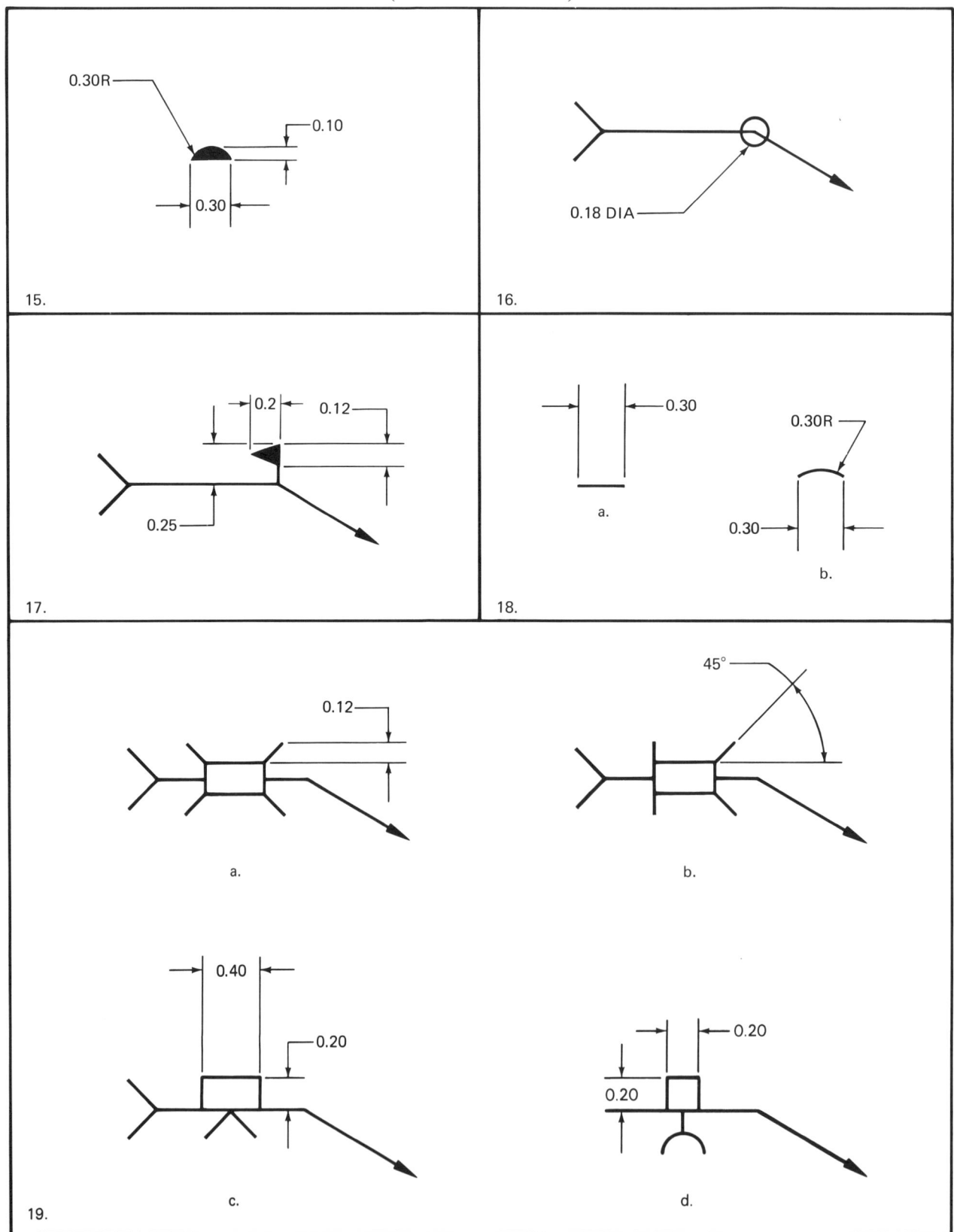

Notes: 1. Unless otherwise specified, tolerance shall be ± .04 or ± 1° as applicable.
2. All radii are minimum dimensions.

Appendix AM (cont.)
Design of Standard Symbols
(Dimensions in millimeters)

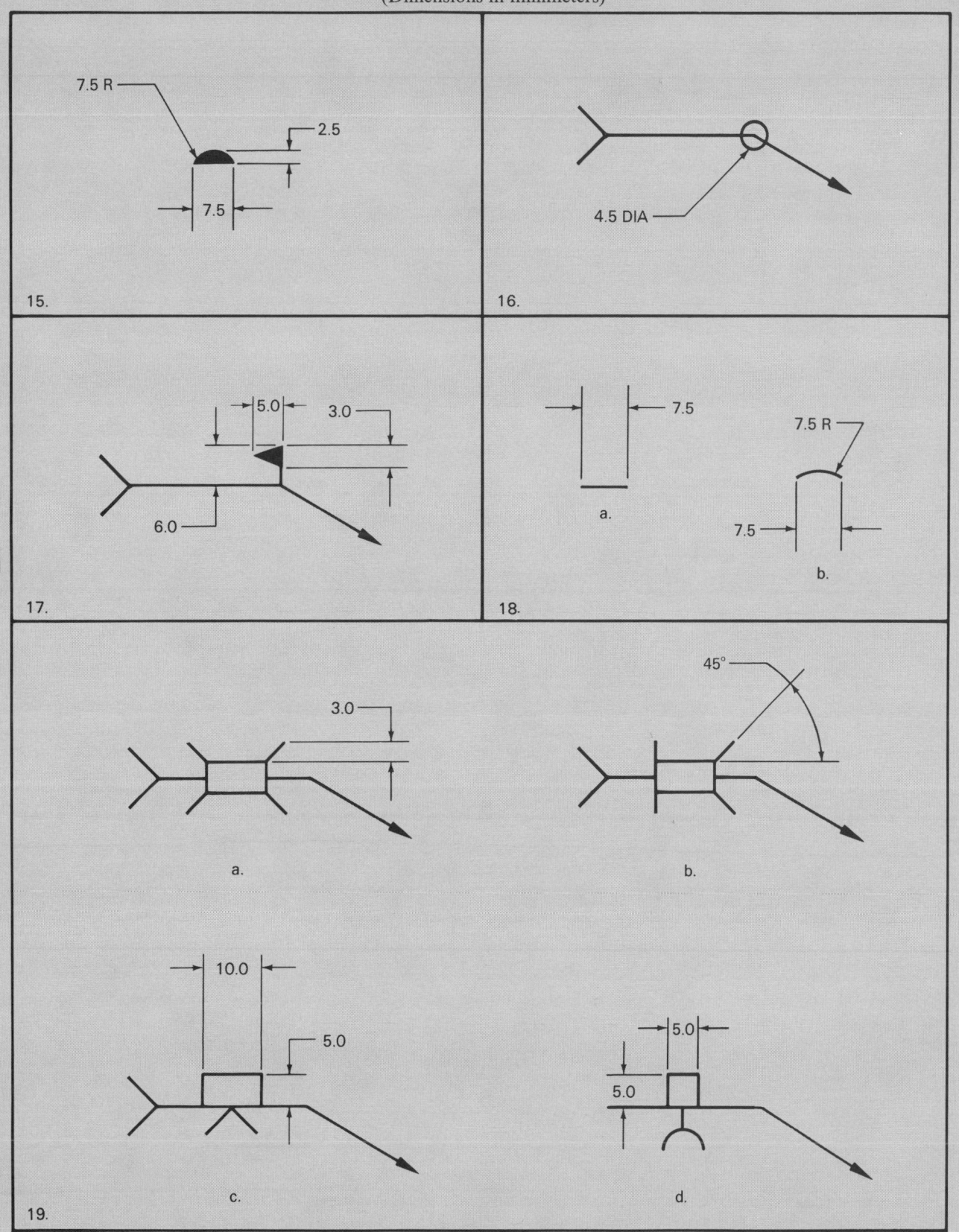

Notes: 1. Unless otherwise specified, tolerances shall be ± 1 or ± 1° as applicable.
2. All radii are minimum dimensions.

Appendix B

Document List

The following is a complete list of the standards prepared by the AWS Committee on Definitions and Symbols:

AWS A2.1-WC	*Welding, Symbols Chart*[1], (Wall size)
AWS A2.1-DC	*Welding, Symbols Chart*[1], (Desk Size)
AWS A2.4	*Standard Symbols for Welding, Brazing and Nondestructive Examination*
AWS A3.0	*Standard Welding Terms and Definitions*, including terms for brazing, soldering, thermal spraying and cutting.

---

[1] A reproduction of the charts is shown on the following two pages. It should be understood that these charts are intended only as shop aids. The only complete and official presentation of the Standard Welding Symbols is in A2.4.

AMERICAN WELDING SOCIETY

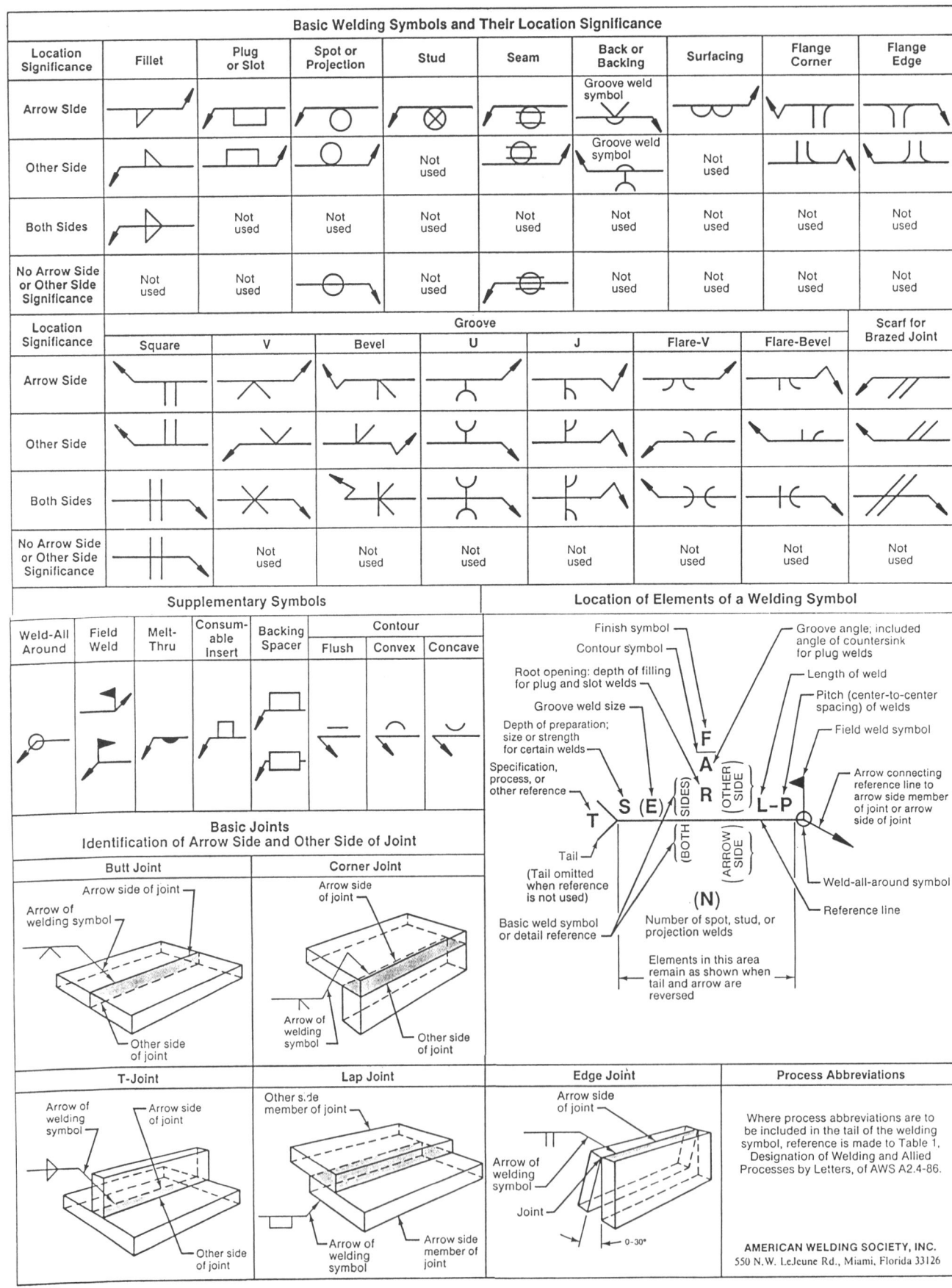

STANDARD WELDING SYMBOLS*

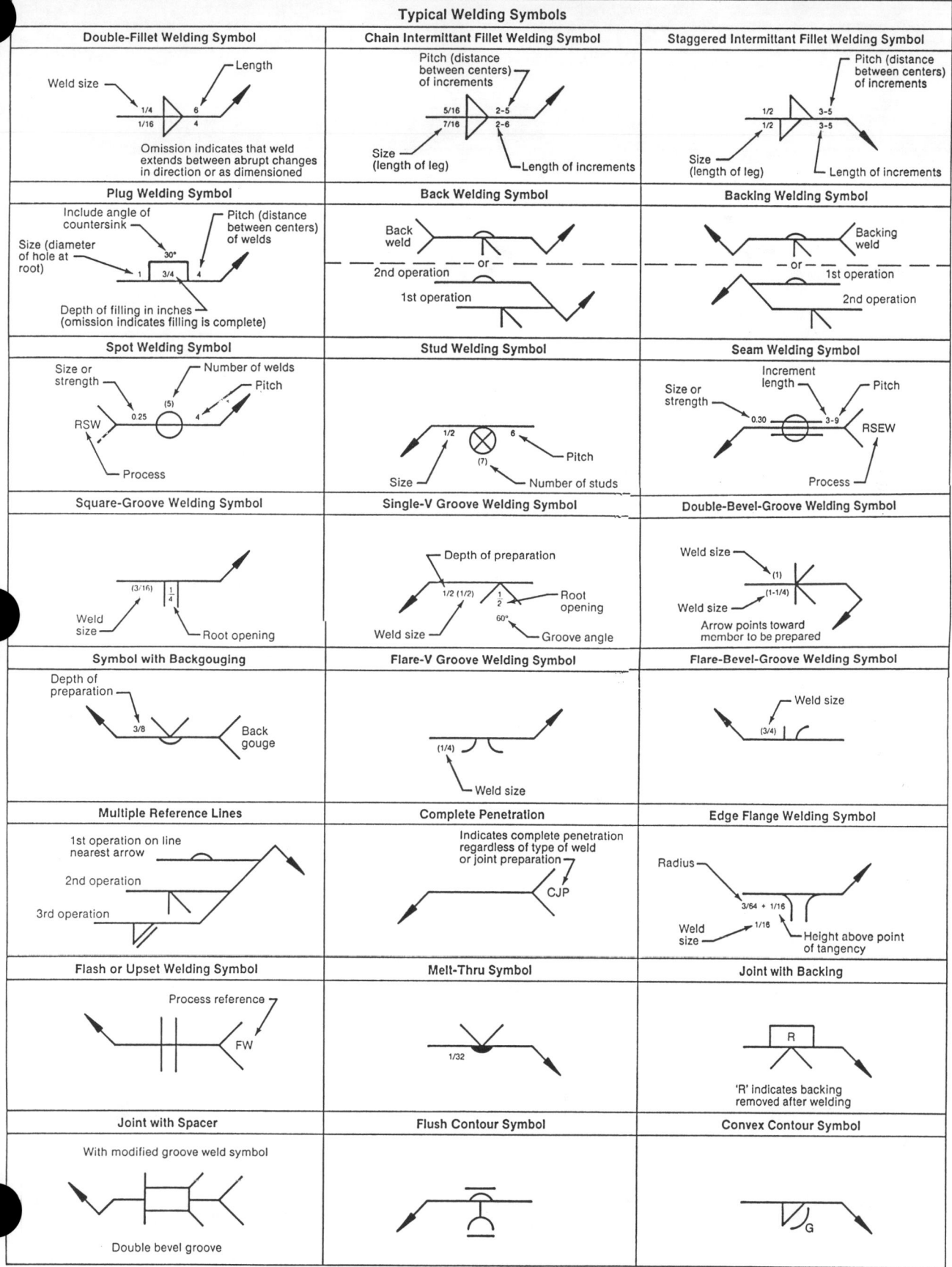

*It should be understood that these charts are intended only as shop aids. The only complete and official presentation of the standard welding symbols is in A2.4.